"Today's final flight out of Afghanistan marks a tragic end to a 20-year conflict. President Biden and his team's disastrous withdrawal—has failed to guarantee America's safety, and worse, left countless Americans, close Afghan partners, and other allies to an unknown and dangerous fate."

—Senator Jim Risch
(Ranking member of Senate Foreign Relations Committee)

"Simply put, when we abandon our friends, our partners around the world start to wonder if they can trust us, if we'll have their backs—this hurts our ability to cooperate with our allies, to deter threats and to provide security for the American people. And it emboldens our adversaries to act more aggressively."

—Senator Roger Wicker
(Comments regarding the withdrawal from Afghanistan)

Left Behind in Afghanistan

A NATIONAL DISGRACE

David Allan Brown

Deeds Publishing | Athens

Copyright © 2022 — David Allan Brown

ALL RIGHTS RESERVED—No part of this book may be reproduced in any form or by any electronic or mechanical means, including information storage and retrieval systems, without permission in writing from the authors, except by a reviewer who may quote brief passages in a review.

Published by Deeds Publishing in Athens, GA
www.deedspublishing.com

Printed in The United States of America

Cover design by Mark Babcock.

ISBN 978-1-950794-86-7

Books are available in quantity for promotional or premium use. For information, email info@deedspublishing.com.

First Edition, 2022

10 9 8 7 6 5 4 3 2 1

Left Behind in Afghanistan

Over 52,000 servicemen and women have been physically injured in recent military conflicts, 500,000 living with invisible wounds, from depression to post-traumatic stress disorder, and another 320,000 have experienced debilitating traumatic brain injuries. Dedicated to the *Wounded Warrior Project* which provides these soldiers a voice, empowers the injured to begin a journey to recovery, and selflessly serves a military population that sacrificed for the benefit and security of our country.

Contents

Prologue	ix
1. "No Man Left Behind Documented in Film	1
2. Portraits of Heroes (No Man Left Behind)	11
3. History of the Region	31
4. President Bush: His Policies and Afghanistan	43
5. President Obama Policies and Afghanistan	47
6. President Trump Policies and Afghanistan	53
7. President Biden Policies & Afghanistan	57
8. President Trump's Deal	61
9. "No Man Left Behind" Policy & Afghanistan	71
10. Methods of Qualifying for Evacuation	83
12. Portrait of an Abandoned Afghan: Massih	93
13. Portrait of an Abandoned Afghan: Ahmed	97
14. Portrait of an abandoned Afghan Khan	99
15. Portrait of an Abandoned Afghan: Romal Noori	107
16. Portrait of an abandoned Afghan: Zak	111
17. Americans Left Behind	117
18. The Fate of the Women of Afghanistan	125
19. The Decision to Abandon Military Bases and Equipment	135
20. Why Not Delay the Withdrawal?	139
Epilogue	151
About the Author	157

Prologue

"We will die if we have to, but we will get him home ... no man left behind."

—Anonymous

"Be what thou wilt; but I will bury him:
Well for me to die in doing that.
I shall rest, a loved one with him whom I have loved, sinless in
My crime: for I owe a longer allegiance to the dead than to the living."

—Antigone determined to recover her brother's body

Throughout history, the United States Military has made a sacred commitment to *Leave No Man Left Behind*.[1] In fact, every branch of the service addresses the concept in their branch-specific ethos. The U.S. Army Ranger creed reads, "I will never leave a fallen comrade to fall into the hands of the enemy..." The Airman's creed, perhaps a little more branch specific, declares, "I will never

---

[1]. It is necessary for readers to understand going forward that the term 'Man' as referred to in the 'No Man Left Behind' ideology refers to all Americans, including our brave women and men in uniform throughout the world.

leave an airman behind," but while referring specifically to airmen, history indicates the Air Force leaves no soldier behind, regardless of military affiliation. The Marines have a motto, "Until they are home, no man left behind" and the Army Soldier's Creed reads, "I will never leave a fallen comrade." These declarations, though each stated slightly different, remain concretely linked in theme: No man should ever be abandoned or left behind. These creeds cannot be found in military manuals or scribed in the Uniform Code of Military Justice, but the ethos is a sacred commitment engrained into the fabric of every American soldier whose actions throughout time clearly reinforces this premise.

The concept is almost as old as warfare itself. The Latin phrase, "Nemo resideo," defined as "Leave No One Behind," dates to the Imperial Roman Army (285-476 A.D.) and remains a sacred belief of the modern United States military. For anyone in combat, this motto is instinctual and uncompromising. Selflessness represents the best in human nature and provides soldiers with confidence and hope that whatever happens out there, they will not be forgotten nor abandoned.

However, recently, not our soldiers, but our country inexplicably abandoned an undetermined number of both Americans and Afghan allies behind enemy lines in Afghanistan. While the true number remains debatable, some suggest upwards of several thousand or more were callously abandoned. Others assert somewhere around two hundred Americans remain precariously stranded in Afghanistan, while an undetermined number of Afghan advisors, humanitarians, and pro-American patriots remain trapped. No matter the number, any figure equal to or greater than one is unacceptable.

Who is responsible for this travesty? Some blame President Biden, some blame then President Trump, while members of

Congress, senators and representatives alike, have spent numerous hours debating this very question on the House floor, to no avail. The answer to the question should not be that difficult to answer—the person in charge is always held accountable for anything that occurs on their watch, while finger pointing accomplishes nothing. Tragically, as the President deflects blame and the Congress debates, our fellow Americans and allies are being hunted, imprisoned, and murdered at the hands of the Taliban and extremists, some at this very moment.

Some may ask, what does it matter what happened in Afghanistan? It matters everywhere that the U.S. attempts to bring peace or positively influence change. It matters to every place that the U.S. tries to "inspire people to take risks." Foreign governments will question future U.S. alliances that are backed by assurances of support, motivated by memories of how America deserted its own people and its allies in Afghanistan.

Is the "No Man Left Behind" ideology something that the Biden administration also holds sacred? If judged by the actions taken in Afghanistan, the answer is doubtful. The Pentagon's top spokesperson said, the Biden administration was "Obviously concerned' for American citizens and Afghan allies still on the ground in Afghanistan following the completion of the U.S. military withdrawal from the country…"[2] Biden acknowledges that those Americans left behind face a threatening future from the Taliban and other extremist groups, but this fact was determined by most even before he made the declaration. Ultimately, the decision to not delay the evacuation until all Americans and allies were out

---

2. Forgey, Quint. "Pentagon: 'We're still obviously concerned' about Americans left in Afghanistan." *Politico*. https://www.politico.com/news/2021/08/31/pentagon-concerned-americans-afghanistan-507955

of the line of fire was reckless and led to a humanitarian crisis of catastrophic proportions.

So, why wasn't the date of the withdrawal postponed until all were evacuated? In the end, the decision to stick to a ceremonial withdrawal date of September 11th despite the overwhelming data suggesting U.S. citizens and their allies would not have time to be evacuated, resulted in one of the most significant foreign policy and humanitarian fiascos in history.

In the meantime, those stranded in Afghanistan in the aftermath of the unnecessarily disastrous withdrawal must not be forgotten. Every effort, every avenue, every tool, every diplomatic solution at the United States' disposal must be employed to rescue those tragically abandoned souls. At the same time, we must wonder, has the recent lack of the Biden Administration's comments on the abandoned Americans and allies serve as a hint that they have already decided to consider them an acceptable sacrifice?

1. "No Man Left Behind" Documented in Film

Military institutions are grounded in law, procedures, and expectations to the point that keeping up with the latest regulations becomes cumbersome and difficult to internalize. For instance, in Army Basic Training, socks, underwear, and t-shirts must be rolled and tucked to meet the designated specifications for locker inspection. A push-up is not a push up unless the back of each arm achieves a ninety-degree (parallel with ground) on the down stroke, while elbows must be locked to finalize each repetition. Unit insignia must be worn one inch above the notches on the collar, with the centerline of the insignia bisecting the notch and parallel to the inside edge of the collar on the Army dress green uniform. These are just a few of the regulations learned early in a soldier's career.

The regulations and directives are exhaustive and impossible to learn in their entirety. While many of these seemingly trivial requirements, insignificant statutes, and bylaws will invariably be forgotten, the longstanding dictum, "No Man Left Behind," resonates throughout American history. There is no compromising or debating that all soldiers, no matter the branch they serve, internalize this belief.

Soldiers have exemplified the 'No Man Left Behind' credence throughout history. Some of the heroic events are documented, while unfortunately many moments fail to be witnessed or those involved died as a result of the act of heroism. Likewise, heroes have emerged from every major American conflict. While these conflicts are deadly, soldiers repeatedly put their life on the line to ensure every man and woman returns home.

Americans enjoy stories of courage and the daunting rescues of seemingly doomed individuals so much that Hollywood has chronicled many military acts of heroism, accentuating awareness of those soldiers who in real-life epitomize the "No Man Left Behind" edict, in film. Word of some soldiers' bravery and courage resonate throughout the American population and as a result, many films document the heroism of soldiers while others go a step further and recognize those who specifically epitomize the "No Man Left Behind" mentality. Some of these nonfiction titles include, *Hacksaw Ridge, Blackhawk Down, Dunkirk* and many others.

Government footage of Hacksaw Ridge

The film *Hacksaw Ridge* documents the actions of **Corporal**

Desmond Doss and his unit, the 2nd Platoon, Company B, 1st Battalion, 307th Infantry, of the 77th Infantry Division. Desmond Doss was born in Lynchburg, Virginia. His mother raised him as a devout Seventh-Day Adventist and instilled nonviolence and Sabbath-keeping in his upbringing. Because of his upbringing, Doss refused to kill an enemy soldier or carry a weapon into combat. He chose military service despite being offered a deferment because of his shipyard work and was viewed as a conscientious objector even after enlisting because of his religious views. He overcame many obstacles to serve and became a combat medic. While serving during the Battle of Okinawa in an area known as Hacksaw Ridge, Doss saved seventy-five wounded comrades by lowering them from a ridge one-by-one to safety below while under the threat of enemy gunfire.

Doss was wounded four times in Okinawa and was evacuated on May 21, 1945, aboard the USS *Mercy*. Doss suffered a left arm fracture from a sniper's bullet while being carried back to Allied lines and at one point had seventeen pieces of shrapnel embedded in his body after an attempt to kick a grenade away from comrades. He was later awarded the Medal of Honor for his actions in Okinawa. Doss refused to evacuate the battlefield even as the enemy advanced on his position. He repeatedly abandoned positions of relative safety, as he stated in an interview with the Medal of Honor Society, to "Get one more," injured/wounded soldier.

Desmond Doss's actions epitomize the principle of No Man Left Behind. On October 12, 1945, President Harry S. Truman presented Doss with the Medal of Honor in a ceremony on the White House lawn. Doss was the first and only conscientious objector to receive the Medal of Honor during World War II. Of the honor, Doss remarked, "I feel that I received the Congressional Medal of Honor because I kept the Golden Rule that we read in

Matthew 7:12. 'All things whatsoever ye would that men should do to you, do ye even so to them.'"

President Harry Truman presenting the Medal of Honor to Private First Class, Desmond Doss

Blackhawk Down, documents the action of units of the 75th Ranger Battalion and the 1st Special Forces Operational Detachment-Delta who deployed to Mogadishu, Somalia in 1993 to capture foreign minister, Omar Salad Elmim and his top political advisor, Mohamed Hassan Awale, during a meeting of rebel leaders. During the mission, rocket-propelled grenades shot down two Blackhawk helicopters. The film documents the efforts to save those perilously pinned down under heavy enemy fire. Two Delta snipers, **Master Sergeant Gary Gordon**, and **Sergeant First Class Randy Shughart**, repeatedly requested to be inserted to help evacuate the injured and wounded. The first two requests were denied, but they were finally granted permission after their third request. They inflicted heavy casualties on the approaching Somali mob. Air support provided fire support for Gordon and Shughart, but an RPG struck the helicopter (Super Six Two). Despite the damage, Super Six-Two managed to land safely. In the ensuing firefight, Gordon was eventually killed. Shughart continued to hold off the mob before he was eventually killed. MSG Gordon and SFC Shughart were posthumously awarded the Medal of Honor. Their actions allowed for wounded to be evacuated and indirectly saved the life of helicopter pilot, Mike Durant.

Master Sergeant Gary Gordon

Sergeant First Class Randall Shughart

The film, *Dunkirk*, chronicles the mass evacuation of hundreds of thousands of French, British, Dutch, Belgian, and Polish troops from the French Port of Dunkirk during World War II. The Battle of Dunkirk between the Allied Forces and Nazi Germany occurred as allies were losing the Battle of France on the Western Front. After a series of attacks, counterattacks, and maneuvers, Allied troops were pinned down with the North Sea to their backs and German forces knocking on the front door making escape nearly impossible. The docks battered by German artillery, ships lined up at the Moles (sea walls protecting the harbor entrance at the east and west borders of the bay), troops endured concentrated German artillery fire and *Luftwaffe* strafing and bombs. The troops fought valiantly providing time for an armada of ships and boats to sweep in and rescue approximately 336,000 troops. While 30-40 thousand troops were eventually captured by the Germans, without the brave soldiers' and civilians' courage to enter the harbor under heavy fire and face a constant threat from the air, the allies avoided an even larger number of casualties.

The Battle of Dunkirk Monument. The French inscription is translated as: "To the glorious memory of the pilots, mariners, and soldiers of the French and Allied armies who sacrificed themselves in the Battle of Dunkirk, May–June 1940." The Dunkirk memorial commemorates the names of the missing & dead of the British Expeditionary Force.

Some involved in the mass evacuation and those actively involved in protecting those while evacuating received various awards, but many accomplishments were not witnessed. In the end, the actions of the hundreds of military personnel and civilians involved in the life-saving mission at Dunkirk exemplify the tenet, "No Man Left Behind."

While Hollywood has done a magnificent job of highlighting some of the real-life soldiers who personify the 'No Man Left Behind' creed, they also capitalize on America's craving for films that depict heroes putting their life on the line for others in fiction. When Forest Gump's unit comes under attack during a well-planned ambush, he repeatedly runs back into the heavy gunfire to save wounded members of his unit. After saving Lieutenant Dan, Forest attempts to go back in while the Lieutenant orders, "I've got an airstrike inbound right now. They're gonna nape the whole area. Gump, you stay right here God damn it. That's an order!" Forest yanks away from the lieutenant's grip and yells, "But I gotta find Bubba!" In perhaps one of the most iconic scenes in any war film, the jungle lights up in a series of explosions behind him as he carries Bubba to safety. The fictional character of Forest Gump serves as a personification of all soldiers throughout time that have risked their lives to rescue their comrades.

Other war films also accomplish the same, 'We all go home' mentality. In *Platoon*, Staff Sergeant Robert Barnes (Tom Berenger) turns on one of his own and shoots Sergeant Elias (Willem Defoe) during a stage of the Tet Offensive. As the evacuation chopper flies away, a member of his squad, Private Chris Taylor (Charlie Sheen) spots Elias running from the jungle with scores of Vietcong soldiers on his tail. The Lieutenant orders, "Get back Get back down there." Even with an overwhelming enemy force in place, the Lieutenant insists on rescuing Elias, but his order is too late.

In *Saving Private Ryan*, a veteran Ranger, Captain John R. Miller (Tom Hanks) is put in command of a squad of men with the single-minded mission of saving one soldier. The soldier, Private First Class James F. Ryan (Matt Damon), must be saved because his three brothers have been killed in action. In the film, Captain Miller and his platoon search for the soldier. Eventually, they find him and send him back home, but at great risk and sacrifice to Captain Miller and his men.

The films *Forest Gump, Platoon,* and *Saving Private Ryan* are just three efforts by Hollywood to chronicle the soldiers' belief that "No Man should be Left Behind." Other films from a myriad of genres do the same by sometimes placing not just soldiers, but civilians and even animated characters in positions to rescue one another. Movies like *Toy Story, Terminator, Guardians of the Galaxy, The Matrix, Captain America,* and many more garner the attention of moviegoers around the world as characters do anything necessary to assist, and in many cases, save their friends and loved ones.

2. Portraits of Heroes (No Man Left Behind)

Privates, seamen, generals, admirals, noncommissioned and commissioned officers; custodians, fast food workers, teachers, laborers, businessmen; Asian, black, white, Indians, and races of all shape, size, and color have served America on the battlefield. Many went on to earn awards and ribbons for their courageous acts and heroics. While many live beyond war to celebrate their victories, receive the accolades of admirers, and go on to live fruitful lives, many died as a result of their valorous deeds. Coming from every walk of life, many originating from other countries, these Americans embody the diverse American Military and in the larger picture, American nationhood.

It would be impossible to say who was the first person to unselfishly risk their life for another on the battlefield, but it is safe to say that most members of the military are willing to risk everything to save their comrades. It would also be a certainty to say that in every war, individuals went to extreme measures in an effort to leave "No Man Left Behind." From World War I to the War in Afghanistan, numerous soldiers have proven that they have internalized and by default exemplified the "No Man Left Behind" truism.

World War I

Sergeant Michael A. Donaldson (January 16, 1884–12 April 1970) served in the United States Army during World War I. He was part of the Irish-American Regiment better known as "The Fighting 69th." While serving in Landres-et St. Georges, France, his regiment entrenched themselves on the crest of a hill after experiencing intense machine gun fire. In the chaotic search for cover, several of their[3] wounded were stranded near enemy lines. Of his own volition, Sergeant Donaldson in daylight and while still under fire, made repeated trips carrying all six wounded men to safety. Donaldson received the Medal of Honor for his actions and later earned numerous awards from France and Montenegro for his actions in St. Georges and later during Champagne-Marne, Aisne-Marne, St. Mihiel, Meuse-Argonne and Defense Sector campaigns.

Sergveant Reidar Waaler (February 12, 1894–5 February 1979) was a Sergeant in the United States Army. Born in Norway, Waaler received the Medal of Honor for his actions during World War I. In the face of heavy artillery and under an onslaught of machine gun fire, Waaler crawled to

---

3. Image of Sergeant Michael A. Donaldson compliments of cmohs.org

a disabled British tank and rescued two men. The tank burned fiercely and yet he returned despite the chance of munitions exploding. He entered the tank to make sure no other occupants remained pinned down. His citation reads:

> "In the face of heavy artillery and machinegun fire, he crawled forward to a burning British tank, in which some of the crew were imprisoned, and succeeded in rescuing two men. Although the tank was then burning fiercely and contained ammunition which was likely to explode at any time, this soldier immediately returned to the tank and, entering it,[4] made a search for the other occupants, remaining until he satisfied himself that there were no more living men in the tank."

World War II

Second Lieutenant, Albert E. Baesel, of the U.S. Army gave his life attempting to rescue a wounded member of his unit during World War II. His citation reads: "Upon hearing that a squad leader of his platoon had been severely wounded while attempting to capture an enemy machine-gun nest about 200 yards in advance of the assault line and somewhat to the right, 2d Lt. Baesel requested permission to go to the rescue of the wounded corporal. After thrice repeating his request and permission having been reluctantly given, due to the heavy

---

4. Image of Sergeant Reidar Waaler compliments of U.S. Military records

artillery, rifle, and machine-gun fire, and heavy deluge of gas in which the company was at the time, accompanied by a volunteer, he worked his way forward, and reaching the wounded man, placed him upon his shoulders and was instantly killed by enemy fire."⁵

Hospital Apprentice First Class, Robert Eugene Bush, served in the U.S Naval Reserve Force and fought during the Okinawa, Jima, and the Ryukyu Islands campaigns. It was during the fighting at the Okinawa, Jima, & Ryukyu Islands on May 2nd, 1945, that Bush, working as a medical corpsman with a rifle company, charged forward, leapfrogging from one injured to the next while under heavy artillery and gun fire. Eventually reaching a critically injured officer, Bush commenced to treating his wounds when the Japanese commenced a ferocious counterattack placing Bush in the midst of the overwhelming Japanese force. Bush although injured himself, continued treating the helpless officer while defending their position with his pistol and later, using an abandoned

---

5. Image of 2nd lieutenant Albert E. Baesel compliments of U.S. Military Records.

carbine. He killed or wounded several enemy soldiers while being critically wounded himself to include the loss of an eye as he defended the wounded soldier and his position until the assault was finally routed. His citation reads as follows:

> "For conspicuous gallantry and intrepidity at the risk of life above and beyond the call of duty while serving as medical corpsman with a rifle company, in action against enemy Japanese forces on Okinawa Jima, Ryukyu Islands, 2 May 1945. Fearlessly braving the fury of artillery, mortar, and machine-gun fire from strongly entrenched hostile positions, Bush constantly and unhesitatingly moved from one casualty to another to attend the wounded falling under the enemy's murderous barrages. As the attack passed over a ridge top, Bush was advancing to administer blood plasma to a Marine officer lying wounded on the skyline when the Japanese launched a savage counterattack. In this perilously exposed position, he resolutely maintained the flow of life-giving plasma. With the bottle held high in one hand, Bush drew his pistol with the other and fired into the enemy ranks until his ammunition was expended. Quickly seizing a discarded carbine, he trained his fire on the Japanese charging point-blank over the hill, accounting for six of the enemy despite his own serious wounds and the loss of one eye suffered during his desperate battle in defense of the helpless man. With the hostile force finally routed, he calmly disregarded his own critical condition to complete his mission, valiantly refusing medical treatment for himself until his officer patient had been evacuated, and collapsing only after attempting to walk to the battle aid station. His daring initiative, great personal valor, and heroic spirit of self-sacrifice in service of others reflect great credit upon Bush and enhance the finest traditions of the U.S. Naval Service."[6]

---

6. Citation and photo provided compliments of the Congressional Medal of Honor Society https://www.cmohs.org/recipients/robert-e-bush

Private First Class Harold Christ Agerholm, served in the U.S. Marine Corp during World War II as a member of the 4th Battalion, 10th Marines, 2nd Marine Division who earned a Medal of Honor serving during the Battle of Saipan in the Marianas Islands. Agerholm volunteered to evacuate wounded during a hostile enemy counterattack. He made several trips into heavy fire to save many individuals until finally being mortally wounded by a sniper. Agerholm's citation reads as follows:

"For conspicuous gallantry and intrepidity at the risk of his life above and beyond the call of duty while serving with the 4th Battalion, 10th Marines, 2d Marine Division, in action against enemy Japanese forces on Saipan, Marianas Islands, 7 July 1944. When the enemy launched a fierce, determined counterattack against our positions and overran a neighboring artillery battalion, Pfc. Agerholm immediately volunteered to assist in the efforts to check the hostile attack and evacuate our wounded. Locating and appropriating an abandoned ambulance jeep, he repeatedly made extremely perilous trips under heavy rifle and mortar fire and single-handedly loaded and evacuated approximately 45 casualties, working tirelessly and with utter disregard for his own safety during a grueling period of more than three hours. Despite intense, persistent enemy fire, he ran out to aid two men whom he believed to be wounded marines, but was himself mortally wounded by a Japanese sniper while carrying out his hazardous mission. Pfc. Agerholm's brilliant initiative, great personal valor, and self-sacrificing efforts in the face of almost certain death reflect the highest credit upon himself and the U.S. Naval Service. He gallantly gave his life for his country."[7]

Korean War

Staff Sergeant Hiroshi Miyamura served in the U.S. Army as part of Company H, 2nd Battalion, 7th Infantry Regiment, 3rd Infantry Division. While under heavy gunfire from enemy soldiers, then Corporal Miyamura, ordered the squad to withdraw while he covered them. Using a machine gun, he thwarted the enemy progress until out of ammunition, then exited his cover using his bayonet in hand-to-hand combat. He again ordered his men to fall back while he continued covering the withdrawal. When last seen, Miyamura's position was overrun while he continued to fight. While Miyamura did not directly rescue any of his comrades, his actions directly saved dozens of lives as they withdrew and exemplify the "No Man Left Behind" mantra. Miyamura's Medal of Honor citation reads as follows:

"Cpl. Miyamura, a member of Company H, distinguished himself by conspicuous gallantry and intrepidity above and beyond the call of duty in action against the enemy. On the night of 24 April, Company H was

---

7. Photo and citation provided by the Congressional Medal of Honor Society https://www.cmohs.org/recipients/harold-c-agerholm

occupying a defensive position when the enemy fanatically attacked, threatening to overrun the position. Cpl. Miyamura, a machine-gun squad leader, aware of the imminent danger to his men, unhesitatingly jumped from his shelter wielding his bayonet in close hand-to-hand combat, killing approximately 10 of the enemy. Returning to his position, he administered first aid to the wounded and directed their evacuation. As another savage assault hit the line, he manned his machine gun and delivered withering fire until his ammunition was expended. He ordered the squad to withdraw while he stayed behind to render the gun inoperative. He then bayoneted his way through infiltrated enemy soldiers to a second gun emplacement and assisted in its operation. When the intensity of the attack necessitated the withdrawal of the company Cpl. Miyamura ordered his men to fall back while he remained to cover their movement. He killed more than 50 of the enemy before his ammunition was depleted and he was severely wounded. He maintained his magnificent stand despite his painful wounds, continuing to repel the attack until his position was overrun. When last seen he was fighting ferociously against an overwhelming number of enemy soldiers. Cpl. Miyamura's indomitable heroism and consummate devotion to duty reflect the utmost glory on himself and uphold the illustrious traditions on the military service."[8]

---

8. Citation and photo provided by the Congressional Medal of Honor Society https://www.cmohs.org/recipients/hiroshi-miyamura

Left Behind in Afghanistan

Master Sergeant William Edward Shuck Jr. was a member of the United States Marine Corp serving in Company G, 3rd Battalion, 7th Marines, 1st Marine Division (REIN) while involved in the Korean War. Shuck assumed command of a rifle squad after his leader was killed in action. He organized two rifle squads and led an assault. Despite being wounded twice, he stayed in a forward position assuring the wounded and the bodies of the dead were evacuated. He earned the Medal of Honor for his actions. His citation reads as follows:

"For conspicuous gallantry and intrepidity at the risk of his life above and beyond the call of duty while serving as a squad leader of Company G, in action against enemy aggressor forces. When his platoon was subjected to a devastating barrage of enemy small-arms, grenade, artillery, and mortar fire during an assault against strongly fortified hill positions well forward of the main line of resistance, S/Sgt. Shuck, although painfully wounded, refused medical attention, and continued to lead his machine-gun squad in the attack. Unhesitatingly assuming command of a rifle squad when the leader became a casualty, he skillfully organized the two squads into an attacking force and led two more daring assaults upon the hostile positions. Wounded a second time, he steadfastly refused evacuation and remained in the foremost position under heavy fire until assured that all dead and wounded were evacuated. Mortally wounded by an enemy sniper bullet while voluntarily assisting in the removal of the last casualty, S/Sgt. Shuck, by his fortitude and great personal valor in the face of overwhelming odds, served to inspire all who observed him. His unyielding courage throughout reflects the highest credit upon himself and the U.S. Naval Service. He gallantly gave his life for his country."[9]

Vietnam

(Image: Clarence Sasser wearing the Medal of Honor [10])

Clarence E. Sasser was drafted into the United States Army in 1967 and by 10 January 1968 he found himself in the heat of battle in Vietnam. While serving as a member of the 9th Infantry Division as a combat medic, Sasser and members of his unit were flown into the Mekong Delta to recon after reports of enemy activity in the area. Approximately a dozen helicopters carried approximately 100 men into the rice paddies where the Vietcong

---

9. Citation and photo provided by the CMOHS. https://www.cmohs.org/recipients/william-e-shuck-jr
10. Photo & citation, compliments of the Congressional Medal OF Honor Society: Lessons of personal Bravery and Self-Sacrifice training materials. Teacher Resource.

had been sighted. The area immediately erupted into a hail of gunfire. Within minutes, thirty men were wounded, and Sasser began hearing the screams, "Medic!" Sasser sprinted across the rice paddies to, as he states, "go to the one who was calling the loudest." He low crawled from casualty to casualty because standing meant certain death from the relentless enemy machine gun and small arms fire. He did his best to keep up with the overwhelming number of wounded and at one point, while dragging a wounded G.I. to cover of an embankment was hit by red-hot shrapnel from an exploding rocket. He pulled the searing metal from his shoulder, refused offers of help from other medics, and rushed back into the rice paddy to help more wounded. He was subsequently wounded in both legs but continued to pull himself along using his arms to assist another soldier. He also directed a group of disoriented soldiers to the cover of a dike so that they could return fire. When interviewed about the dire situation, Sasser replied, "I felt that if I could get the guys up and fighting," he said later, "we might all get out of there somehow." Although faint from loss of blood, Sasser repeatedly rushed into enemy fire to assist and evacuate wounded soldiers. Clarence Sasser's courage under fire and drive to get every man out epitomizes the Leave No Man Left Behind mantra. His Medal of Honor was presented to him by then President Richard M. Nixon on March 7, 1969 and reads:

> For conspicuous gallantry and intrepidity in action at the risk of his life above and beyond the call of duty. Specialist 5th Class Sasser distinguished himself while assigned to Headquarters and Headquarters Company, 3d Battalion. He was serving as a medical aidman with Company A, 3d Battalion, on a reconnaissance in force operation. His company was making an air assault when suddenly it was taken under heavy small arms, recoilless rifle, machinegun and rocket fire

from well-fortified enemy positions on three sides of the landing zone. During the first few minutes, over 30 casualties were sustained. Without hesitation, Specialist 5th Class Sasser ran across an open rice paddy through a hail of fire to assist the wounded. After helping one man to safety, was painfully wounded in the left shoulder by fragments of an exploding rocket. Refusing medical attention, he ran through a barrage of rocket and automatic weapons fire to aid casualties of the initial attack and, after giving them urgently needed treatment, continued to search for other wounded. Despite two additional wounds immobilizing his legs, he dragged himself through the mud toward another soldier 100 meters away. Although in agonizing pain and faint from loss of blood, Specialist 5th Class Sasser reached the man, treated him, and proceeded on to encourage another group of soldiers to crawl 200 meters to relative safety. There he attended their wounds for five hours until they were evacuated. Specialist 5th Class Sasser's extraordinary heroism is in keeping with the highest traditions of the military service and reflects great credit upon himself, his unit, and the U.S. Army.

Sammy L. Davis enlisted in the Army straight out of high school in 1966. He was assigned to the 4th Artillery. In November of 1967, his unit was flown into an area west of Cai Lay to set up a forward fire support base. Early the next morning, just after midnight, his unit came under attack by approximately 1,500 Vietcong soldiers. Davis manned a machine gun and provided cover for the Howitzer gun crew. Davis was knocked from his position by an explosion, knocking him unconscious. He discovered he was seriously wounded when he woke. Realizing that his unit was going to be overrun, Davis fired a[11] nearby artillery piece at point blank range, eliminating many of the enemy but shortly after, a mortar round exploded nearby taking him from his feet. Once he recovered, he heard voices of other soldiers cut off by a nearby river. He managed to get across the river, found three wounded soldiers, gave two of them morphine and evacuated the most seriously wounded soldier. He then went back and evacuated the other two. Eventually, he made his way to another Howitzer group and resumed fighting. Davis's selfless act saved three soldiers that day because he refused to let the enemy kill or capture his comrades; he refused to leave his fellow soldiers behind. On November 19, 1968, President Lyndon B. Johnson presented the Medal of Honor to Sammy Davis for his heroic actions. His citation reads:

> For conspicuous gallantry and intrepidity in action at the risk of his life above and beyond the call of duty. Sgt. Davis (then Pfc.) distinguished himself during the early morning hours while serving as a cannoneer with Battery C, at a remote fire support base. At approximately 0200 hours the fire support base was under heavy enemy mortar attack. Simultaneously, an estimated reinforced Viet Cong battalion launched a fierce ground assault upon the fire support base. The attacking enemy drove to within 25 meters of the friendly positions. Only a river separated the Viet Cong from the fire support base. Detecting a nearby enemy position, Sgt. Davis seized a machine gun and provided covering fire for his gun crew, as they attempted to bring direct artillery fire on the enemy. Despite his efforts, an enemy recoilless-rifle round scored a direct hit upon the artillery piece. The resultant blast hurled the gun crew

---

11. Photo & citation, compliments of the Congressional Medal OF Honor Society: Lessons of personal Bravery and Self-Sacrifice training materials. Teacher Resource.

from their weapon and blew Sgt. Davis into a foxhole. He struggled to his feet and returned to the howitzer, which was burning furiously. Ignoring repeated warnings to seek cover, Sgt. Davis rammed a shell into the gun. Disregarding a withering hail of enemy fire directed against his position, he aimed and fired the howitzer which rolled backward, knocking Sgt. Davis violently to the ground. Undaunted, he returned to the weapon to fire again when an enemy mortar round exploded within 20 meters of his position, injuring him painfully. Nevertheless, Sgt. Davis loaded the artillery piece, aimed, and fired. Again he was knocked down by the recoil. In complete disregard for his safety, Sgt. Davis loaded and fired three more shells into the enemy. Disregarding his extensive injuries and his inability to swim, Sgt. Davis picked up an air mattress and struck out across the deep river to rescue three wounded comrades on the far side. Upon reaching the three wounded men, he stood upright and fired into the dense vegetation to prevent the Viet Cong from advancing. While the most seriously wounded soldier was helped across the river, Sgt. Davis protected the two remaining casualties until he could pull them across the river to the fire support base. Though suffering from painful wounds, he refused medical attention, joining another howitzer crew which fired at the large Viet Cong force until it broke contact and fled. Sgt. Davis' extraordinary heroism, at the risk of his life, is in keeping with the highest traditions of the military service and reflect great credit upon himself and the U.S. Army.

War on Terror in Afghanistan

Staff Sergeant Salvatore Giunta served in the United States Army. While assigned to Company B, 2nd Battalion (Airborne), 503rd Infantry Regiment, 173rd Airborne Brigade Combat Team, in the Korengal Valley, he distinguished himself by rescuing many members of his squad to include his wounded squad leader, and other members of the squad. While on patrol, his unit was ambushed. He pushed forward through heavy fire and witnessed two insurgents carrying a wounded squad member away. Firing on the enemy, he was able to recover the critically wounded soldier until others arrived to provide cover fire. Sergeant Guinta's Medal of Honor citation reads as follows:

"Specialist Salvatore A. Giunta distinguished himself conspicuously by gallantry and intrepidity at the risk of his life above and beyond the call of duty in action with an armed enemy in the Korengal Valley, Afghanistan, on October 25, 2007. While conducting a patrol as team leader with Company B, 2d Battalion (Airborne), 503d Infantry Regiment, Specialist Giunta and his team were navigating through harsh terrain when they were ambushed by a well-armed and well-coordinated insurgent force. While under heavy enemy fire, Specialist Giunta immediately sprinted towards cover and engaged the enemy. Seeing that his squad leader had fallen and believing that he had been injured, Specialist Giunta exposed himself to withering enemy fire and raced towards his squad leader, helped him to cover, and administered medical aid. While administering first aid, enemy fire struck Specialist Giunta's body armor and his secondary weapon. Without regard to the ongoing fire, Specialist Giunta engaged the enemy before prepping and throwing grenades, using the explosions for cover in order to conceal his position. Attempting to reach additional wounded fellow soldiers who were separated from the squad, Specialist Giunta and his team encountered a barrage of enemy fire that forced them to the ground. The team continued forward and upon reaching the wounded soldiers, Specialist Giunta realized that another soldier was still separated from the element. Specialist Giunta then advanced forward on his own initiative. As he crested the top of a hill, he observed two insurgents carrying away an American soldier. He immediately engaged the enemy, killing one and wounding the other. Upon reaching the wounded soldier, he began to provide medical aid, as his squad caught up and provided security. Specialist Giunta's unwavering courage, selflessness, and decisive leadership while under extreme enemy fire were integral to his platoon's ability to

Left Behind in Afghanistan

defeat an enemy ambush and recover a fellow American soldier from the enemy. Specialist Salvatore A. Giunta's extraordinary heroism and selflessness above and beyond the call of duty are in keeping with the highest traditions of military service and reflect great credit upon himself, Company B, 2d Battalion (Airborne), 503d Infantry Regiment, and the United States Army."[12]

Staff Sergeant Clinton Romesha served in the United States Army. While serving in the Outpost Keating, Kamdesh District, Nuristan Province, Afghanistan, as a member of Bravo Troop, 3rd Squadron, 61st Calvary regiment, 4th Brigade Combat Team, 4th Infantry Division, Romesha while under attack from a large enemy force pushed forward with other members of his unit to recover the bodies of fallen comrades from enemy forces. His Medal of Honor citation reads:

"For conspicuous gallantry and intrepidity in action at the risk of his life above and beyond the call of duty while serving as a Section Leader with Bravo Troop, 3d Squadron, 61st Cavalry Regiment, 4th Brigade Combat Team, 4th Infantry Division, during combat operations against an armed enemy at Combat Outpost Keating, Kamdesh District, Nuristan Province, Afghanistan on 3 October 2009. On that

---

12. Citation and photo provided by the Congressional Medal of Honor Society https://www.cmohs.org/recipients/salvatore-a-giunta

morning, Staff Sergeant Romesha and his comrades awakened to an attack by an estimated 300 enemy fighters occupying the high ground on all four sides of the complex, employing concentrated fire from recoilless rifles, rocket propelled grenades, anti-aircraft machine guns, mortars, and small arms fire. Staff Sergeant Romesha moved uncovered under intense enemy fire to conduct a reconnaissance of the battlefield and seek reinforcements from the barracks before returning to action with the support of an assistant gunner. Staff Sergeant Romesha took out an enemy machine gun team and, while engaging a second, the generator he was using for cover was struck by a rocket-propelled grenade, inflicting him with shrapnel wounds. Undeterred by his injuries, Staff Sergeant Romesha continued to fight and upon the arrival of another soldier to aid him and the assistant gunner, he again rushed through the exposed avenue to assemble additional soldiers. Staff Sergeant Romesha then mobilized a five-man team and returned to the fight equipped with a sniper rifle. With complete disregard for his own safety, Staff Sergeant Romesha continually exposed himself to heavy enemy fire, as he moved confidently about the battlefield engaging and destroying multiple enemy targets, including three Taliban fighters who had breached the combat outpost's perimeter. While orchestrating a successful plan to secure and reinforce key points of the battlefield, Staff Sergeant Romesha maintained radio communication with the tactical operations center. As the enemy forces attacked with even greater ferocity, unleashing a barrage of rocket-propelled grenades and recoilless rifle rounds, Staff Sergeant Romesha identified the point of attack and directed air support to destroy over 30 enemy fighters. After receiving reports that seriously injured soldiers were at a distant battle position, Staff Sergeant Romesha and his team provided covering fire to allow the injured soldiers to safely reach the aid station. Upon receipt of orders to proceed to the next objective, his team pushed forward 100 meters under overwhelming enemy fire to recover and prevent the enemy fighters from taking the bodies of the fallen comrades. Staff Sergeant Romesha's heroic actions throughout the day-long battle were critical in suppressing an enemy that had far greater numbers. His extraordinary efforts gave Bravo Troop the opportunity to regroup, reorganize, and prepare for the counterattack that allowed the Troop to account for its personnel and secure Combat Post Keating. Staff Sergeant Romesha's discipline and extraordinary heroism above and beyond the call of duty reflect great credit upon himself, Bravo Troop, 3d Squadron, 61st Cavalry Regiment, 4th Brigade Combat Team, 4th Infantry Division and the United States Army." [13]

---

13. Citation and photo provided by the Congressional Medal of Honor Society https://www.cmohs.org/recipients/clinton-l-romesha

3. History of the Region

Throughout time, Afghanistan has seemingly always been at war, but a cursory overview of the last five decades will provide ample background leading up to the War on Terror and the eventual withdrawal of American Forces in 2021. On July 17, 1973, an Afghan coup détat was led by Army General and Prince Mohammed Daoud Khan against his cousin, King Mohammed Zahir Shah. Khan established the People's Democratic Party of Afghanistan and named himself the president. Two years later, between 1975 and 1977, Khan proposed a new Constitution with two key efforts; first, to grant women more rights and second, to modernize the largely Communist state.

A year later, in 1973, King Mohammed Zahir Shah was forced out of power by a Communist coup that ends with Nur Mohammad Taraki (A founding member of the Communist Party) assuming the presidency while Babrak Karmel is named his Deputy Prime Minister.

At this point, it is important to understand the Soviet Union's influence in the region. In 1953, then Afghan General, Mohammed Daoud Khan, cousin of the king, became prime minister. He had close ties with the Union of Soviet Socialist Republic (USSR) and leaned on them for economic and military assistance.

Simultaneously, he put into place many social reforms. Three years later, in 1956, Soviet leader, Nikita Khrushchev agreed to provide assistance, and the two countries became close allies. Soviet influence remained strongly entrenched as they backed the Afghan government and leveraged their presence in the region. Soviet influence continued throughout the 1950s into the 1980s.

In 1978, President Nur Mohammad Taraki and Prime Minister Babrak Karmel proclaimed independence from Soviet influence, and declared their policies be based on Afghan nationalism and socioeconomic justice, while primarily enforcing a strict return to Islamic principles. Taraki eventually signed a friendship treaty with the Soviet Union, but a rivalry between Taraki and Hafizullah Amin (another influential communist leader), led to fighting between the two sides. Simultaneously, ethnic and conservative Islamic leaders began an armed revolt resulting from Khans proposed social changes. In June of the same year, the Mujahadeen guerrilla movement formed to fight against the Soviet-backed government.

The United States quietly provided support for the Afghan government throughout the late seventies, but abruptly cut off assistance after American Ambassador Adolph Dubs was killed. Soon after, a power struggle began between Taraki and Deputy Prime Minister Hafizullah Amin. On September 14, 1979, Taraki was killed in a confrontation with Amin supporters.

On December 4, 1979, the Soviet Union (USSR) invaded Afghanistan to assist the struggling Communist regime. Amin and many of his followers were executed shortly after the invasion, resulting in Deputy Prime Minister Babrak Karmel being named the Prime Minister. Violent public demonstrations led by disgruntled citizens soon became widespread. The Mujahadeen rebels united against the Soviet backed army beginning in the early 1980s.

By 1982, the result of fierce fighting by both sides created a humanitarian crisis in which nearly three million citizens fled to Pakistan and another one and a half million fled to Iran. After fierce fighting, the Soviets gained control of most of the urban areas while Afghan rebels held control of most suburban locations.

In 1984, Saudi born Osama bin Laden made his first trip to Afghanistan to assist anti-Soviet forces. Two years later, the Mujahadeen began receiving arms from the United States, China, and England through Pakistan. Four years later, Osama Bin Laden and a group of influential Islamists formed the group al-Qaida to continue to resist Soviet forces. Bin Laden then went a step further to proclaim a jihad, or holy war not just against the Soviets, but also against any other nation that opposed their goal of a "Pure Islamic nation." Soon after, al-Qaida claimed a victory over the Soviets, and they quickly turned their focus to America by declaring the United States the main obstacle to al-Qaida establishing an Islamic State.

Five years later, on April 14, 1988, the United States, the Soviet Union, Pakistan, and Afghanistan met at Geneva to sign a peace accord. The Geneva accords accomplished many things to include guaranteed Afghan independence and set a timeline for the withdrawal of over 100,000 Soviet troops. The Mujahadeen were not included in the peace discussions, nor were they invited to the peace accords. As a result, they continued resisting the leadership of Afghan President Dr. Mohammad Najibullah (who the Mujahadeen claimed was a Puppet Soviet influenced president). The Mujahadeen elect Sibhatullah Mojadidi as the head of their exiled force. Eventually, in 1992, rebels along with disgruntled members of the Afghan Army forced Najibullah out of office. Guerrilla leader, Ahmad Dhah Masood led troops into Kabul (the capital). The United Nations offered protection to Najibullah. The

Mujahadeen, while in turmoil since various warlords had different ideas for the future of Afghanistan, formed a largely Islamic State with Professor Burhannudin Rabbani as president.

Around 1994-95, the Mujahadeen guerrilla forces formed an Islamic militia called the Taliban. They promised the people of Afghanistan peace. After so many years of warring, famine, and drought, the people approved of the Taliban and their promise to uphold tradition. The Taliban went on to crack down on crime, and outlaw the cultivation of poppies for the opium trade, as well as severely restricted the education and employment of women. They mandated that all women be fully veiled, and they were no longer permitted outside alone. The Taliban began public executions and amputations to enforce their version of Islamic law. The United States refused to recognize the Taliban as a governing state.

In 1997, the Taliban executed Najibullah while ethnic groups from the north and the south under Masood, aided in part by Hamid Karzai, battled the Taliban for control. Months later, in early 1998, al-Qaida bombed two American embassies in Africa. President Clinton ordered the bombing of Bin Laden's training camps in Afghanistan in retaliation. The bombings weren't effective in that they missed the leadership of the terrorist group which were the primary targets.

The United States demanded that Bin Laden (now viewed as an international terrorist) be extradited to stand trial for the bombings in Africa, but the Taliban refused to do so. In turn, the United Nations used sanctions as a tool for the Taliban's unwillingness to extradite Bin Laden (restricting trade and economic support).

The Taliban, under the leadership of Bin Laden, continued to terrorize the people of their own nation as well as backed acts of terrorism throughout the world. Some of these atrocities included mass killings of civilians, burning villages and orchards, torturing detainees, displacing civilians by force, and rejecting international pleas that the Taliban not blow up two 1,500-year-old colossal Buddha statues carved into a mountainside in Bamiyan, Afghanistan. The Taliban said that the statues were "idols" prohibited under Islam. A month later, on September 4, 2001, the Taliban arrested eight international aid workers (to include Americans) and sixteen Afghan aid workers. According to a CNN report from the same date, "The act of trying to convert Muslims to Christianity is considered illegal under the strict Islamic law enforced by the ruling Taliban."

The act is punishable by death. The individuals were to be tried in Morafia High Court. According to a Reuters, "The chief justice vowed the trial, which would be carried out in the light of Islamic Sharia law, would be fair and the accused would be given the chance to speak and 'defend themselves without any compulsion or fear.'" The workers were held in various Afghan prisons for months but under increasing pressure from the international community, the foreign detainees were finally released on November 15, 2001 but the Afghan Muslims were sentenced to death.

On September 11, 2001, in perhaps one of the boldest terrorist attacks in modern history, nineteen militant hijackers associated with al Qaida commandeered four commercial airliners and carried out suicide attacks against a number of targets in the United States. One plane hit the Pentagon in Arlington, Virginia, two flew into the Twin Towers of the World Trade Center complex, and the fourth crashed into a field in Shanksville, Pennsylvania.

David Allan Brown

A picture of the Twin Towers after terrorists hijacked and flew a commercial airliner into both the north and south towers.[14]

The attacks devastated the American people but at the same time brought a divided United States together. 2,996 people died as a result of the 9/11 terrorist attacks. This number includes 2,763 at the World Trade Center (includes 343 fire fighters, 23 New York City police officers, 37 Port Authority officers), 189 at the Pentagon (64 from flight 77 that struck the building), and 44 heroes aboard Flight 93 who thwarted terrorist attempts to inflict greater damage when the plane crashed in a field in Pennsylvania. U.S. officials determined Osama Bin Laden to be the prime suspect of the attack. The attack led to the death of 2,996 innocent people nationwide and to 2,763 at the World Trade Center.

---

14. Clark, Roger (Photographer -*Associated Press*). Photo in the *L.A. Times*. "World Trade Center and Pentagon attacked on September 11th 2001." By Matea Gold & Maggie Farley. 12 September 2001. https://www.latimes.com/travel/la-xpm-2001-sep-12-na-sept-11-attack-201105-01-story.html

The United States and the International community demanded that the Taliban turn over Bin Laden to face charges for the act of terrorism, but they refused to do so. In response, the United States and Britain launched a series of airstrikes aimed at Taliban targets and bases thought to be part of the al-Qaida network. The Afghan Northern Alliance simultaneously engaged in intense battle with the Taliban and entered Kabul. The Taliban retreated southward to Kandahar. Months later, the Taliban abandoned Kandahar as their hold on Afghanistan dwindled. Finally, in late 2001, the Taliban were forced out of their last Afghan territory of Zabul, and it was reported in the Pakistan-based Islamic Press that "the rule of the Taliban in Afghanistan has totally ended." Hamid Karzai assumed the leadership position in the interim government in Afghanistan (he was in exile in Pakistan for eight years). Karzai had the support of the United States. The United Nations sponsored a conference to establish an interim government and Karzai officially became the leader of the six-month interim government. Karzai appointed the members of his government who served until 2004 when elections were held.

Amid continued violence in Kabul, NATO sent in an international force to provide security in the region. The Afghan council adopted a new Constitution after taking input from nearly 500,000 Afghan citizens. The new Constitution established the office of the president and two vice presidents. The document also called for equality for women. In October of 2004, elections were held. More than ten and a half million Afghans registered to vote. Karzai was re-elected as president, garnering 55 percent of the vote among eighteen candidates. The next year, the country's first parliamentary elections in over thirty years were held and they met for the first time in December of the same year. At the same time, Taliban and al-Qaida forces continued to battle Afghan national forces, which resulted in NATO expanding its peacekeeping operations to the more southerly portion of Afghanistan. NATO forces were subjected to increasing suicide bombings and raids against the international force. Taliban commander Mullah Dadullah was killed in a US- led covert operation in Afghanistan.

In 2008, a donor's conference in Paris raised over fifteen billion dollars in aid to Afghanistan. President Hamid Karzai made a commitment to fight continued corruption in his government as a result.

On March 27, 2009, President Barack Obama announced a comprehensive, new strategy for Afghanistan and Pakistan. In his speech to the nation, President Obama declared:

> "Many people in the United States—and many in partner countries that have sacrificed so much—have a simple question: What is our purpose in Afghanistan? After so many years, they ask, why do our men and women still fight and die there? And they deserve a straightforward answer.
>
> "So let me be clear: Al Qaeda and its allies—the terrorists who planned and supported the 9/11 attacks—are in Pakistan and Afghanistan. Multiple intelligence estimates have warned that al-Qaeda is actively planning attacks on the United States homeland from its safe haven in Pakistan. And if the Afghan government falls to the Taliban—or allows al Qaeda to go unchallenged—that country will again be a base for terrorists who want to kill as many of our people as they possibly can…the future of Afghanistan is inextricably linked to the future of its neighbor, Pakistan.
>
> "In the nearly eight years since 9/11, al Qaeda and its extremist allies have moved across the border to the remote areas of the Pakistani frontier. This almost certainly includes al Qaeda's leadership: Osama bin Laden and Ayman al-Zawahiri. They have used this mountainous

terrain as a safe haven to hide, to train terrorists, to communicate with followers, to plot attacks, and to send fighters to support the insurgency in Afghanistan. For the American people, this border region has become the most dangerous place in the world. But this is not simply an American problem — far from it. It is, instead, an international security challenge of the highest order."

President Obama went on to discuss the presence of al-Qaeda and Taliban forces within the Pakistani borders. As a result, President Obama appointed Richard Holbrooke as a special envoy to Afghanistan and Pakistan and dispatched approximately 17,000 more combat troops in 2009. Assistance also included training and support for the Pakistani military. Later, in 2010, President Obama appointed General David Petraeus, head of the U.S. Central Command.

Then on May 2, 2011, U.S. forces raided Abbottabad, Pakistan and killed Osama Bin Laden, mastermind of the September 11, 2011, terrorist attacks on the United States. The fifty-four-year-old leader of al-Qaeda was one of Americas most wanted for nearly a decade-long international hunt.

The following is a synopsis of the raid of Bin Laden's compound as provided by The HISTORY® Channel[15].

The raid began around 1 a.m. local time (4 p.m. EST on May 1, 2011 in the United States), when 23 U.S. Navy SEALs in two Black Hawk helicopters descended on the compound in Abbottabad, a tourist and

---

15. Report provided by History.com editors and published on: https://www.history.com/this-day-in-history/osama-bin-laden-killed-by-u-s-forces

military center north of Pakistan's capital, Islamabad. One of the helicopters crash-landed into the compound, but no one aboard was hurt. During the raid, which lasted approximately 40 minutes, five people, including bin Laden and one of his adult sons, were killed by U.S. gunfire. No Americans were injured in the assault. Afterward, bin Laden's body was flown by helicopter to Afghanistan for official identification, then buried at an undisclosed location in the Arabian Sea less than 24 hours after his death, in accordance with Islamic practice.

Just after 11:30 p.m. EST on May 1 (Pakistan's time zone is nine hours ahead of Washington, D.C.), President Barack Obama, who monitored the raid in real time via footage shot by a drone flying high above Abbottabad, made a televised address from the White House, announcing bin Laden's death. "Justice has been done," the president said. After hearing the news, cheering crowds gathered outside the White House and in New York City's Times Square and the Ground Zero site.

The success of the raid received international attention and provided a much-needed victory for the United States military and the intelligence community.

Just two years later, the Afghan Army assumed all military and security operations from NATO. This was a huge step for the Afghan government, as the move tested the readiness and strength of their military forces. In response to the announcement, President Obama announced a timetable for reducing the number of the U.S. troop size in Afghanistan. The plan included a methodical reduction of troops between 2014 and 2016, but after deteriorating conditions in Afghanistan, President Obama abandoned his plan, leaving 5,500 troops to train and advise the Afghan military while NATO officially ending its combat involvement. During this period, Ashraf Ghani was elected president (2014).

Once Donald Trump was elected president, he declared that

there would be a continued American presence in Afghanistan to prevent a "vacuum for terrorists" if America were to completely withdraw troops. Two years later, President Trump and the Taliban signed a deal that established the preliminary terms of a U.S. withdrawal by May 2021. In November 2020, the United States announced plans to cut U.S. troop size to approximately 2,500 personnel shortly before the inauguration of Joseph Biden. Once elected, President Biden announced his intention to complete U.S. troop withdrawal in accordance with the deal President Trump marshalled; however, he delayed the withdrawal and eventually committed to the mainly ceremonial withdraw date of 11 September 2021. A series of events quickly occurred to include:

a. July 5, 2021: The United States left Bagram Airfield without informing the Afghan commander.
b. U.S. announced a Taliban takeover was not inevitable following the withdrawal of American troops.
c. The Afghan government collapsed in response to the influx of Taliban fighters.
d. The Taliban took control of Kabul
e. Suicide bombings occurred outside of the Kabul Airport as thousands of Afghan citizens attempted to flee. Approximately 169 Afghan citizens were killed and thirteen U.S. Marines.
f. ISIS-K[16] (Khorasan) claimed responsibility for the bombings
g. Despite the bombings and imminent proof that the

---

16. The extremist group ISIS-K, the affiliate of the terror group ISIS, which uses the "K" to reference an old name for Afghanistan, Khorasan. The group first appeared in Eastern Afghanistan in late 2014.

Taliban will/have assumed control in Afghanistan, President Joe Biden continued the withdrawal.

h. President Biden vowed to retaliate against the perpetrators of the attack when he states, "We will not forgive. We will not forget. We will hunt you down to make you pay."

i. The final contingent of U.S. troops from Afghanistan departed from the Kabul Airport on August 31, 2021[17].

j. In August 2021, the Pentagon reported that some Americans were unable to evacuate and would have to rely on diplomatic channels to exit.

---

17. This officially ends America's war in Afghanistan.

4. President Bush: His Policies and Afghanistan

"The attack took place on American soil, but it was an attack on the heart and soul of the civilized world. And the world has come together to fight a new and different war, the first, and we hope the only one, of the 21st century. A war against all those who seek to export terror, and a war against those governments that support or shelter them."

—**President George W. Bush**, October 11, 2001

On September 18, 2001, President George W. Bush signed into law a joint resolution authorizing the use of force against those responsible for attacking the United States on 9/11. This joint resolution was later cited by the Bush administration as legal rationale for its decision to take "sweeping measures to combat terrorism…"[18] Shortly after, on October 7, 2001, the United States

---

18. https://www.cfr.org/timeline/us-war-afghanistan

launched "Operation Enduring Freedom." Likewise, the world responded with an unprecedented coalition against international terrorism in response to the September 11, 2001, attack. President Bush [immediately] implemented a comprehensive and visionary foreign policy against international terrorism. The president's policy put "The world on notice that any nation that harbors or supports terrorism will be regarded as a hostile regime."[19] With that, the War on Terror began on September 11, 2001, when members of al-Qaida hijacked four commercial airliners and intentionally slammed into predetermined targets. One hit the Pentagon, two struck the Twin Towers, and efforts to hit an unknown fourth target were thwarted by the heroic efforts of passengers aboard United Airlines Flight 93. The plane crashed in a field in Somerset, Pennsylvania. Once these events occurred, the War on Terror officially commenced.

Because of the September 11, 2001, attacks, then President George Bush officially commenced America's "War on Terror" and later added the Moniker "Global." His first step was to form a coalition of other nations. With the full support of the Congress, President Bush ordered the American military to establish al-Qaida targets consisting of their strongholds, training bases, and any elements of the Taliban government that sheltered or sponsored al-Qaida sponsored activities.[20]

He did not act without first receiving Congressional approval who quickly identified the enemy as all "nations, organizations, or persons" responsible for the September 11 attacks. The

---

19. U.S. Department of State https://georgewbush-whitehouse.archives.gov/infocus/achievement/chap1-nrn.html
20. Kenneth Katzman & Clayton Thomas, Cong. Research Service, RL305088, Afghanistan: Post-Taliban Governance, Security, and U.S. policy (2017)

authorization for use of military force provided full authority for President Bush to use armed forces against those "he determines, planned, authorized, committed, or aided the terrorist attacks'"[21] against the United States.

According to Lieutenant (U.S. Army retired) Colonel Jeffrey F. Addicot[22], the military campaign to oust the Taliban and close the al-Qa'ida training camps took only three months, October to December 2001. He went on to discuss how combat hostilities continued throughout both terms of the Bush presidency (against the Taliban and al-Qaida).[23] Osama Bin Laden fled and remained in hiding for approximately ten years following the offensive. He remained on the run until then President Barack Obama authorized a secret mission on the early hours of (1:00 A.M. Pakistan Standard Time) May 1, 2011, where 23 Navy Seals raided Bin Laden's Abbottabad Compound in Pakistan killing five terrorists to include Osama Bin Laden and his son.

The next area of concern became weapons of mass destruction and the idea of "rogue nations who posed a direct threat to the United States by possessing or seeking to possess WMDs"[24] and the idea of them being used in a new terror strike against the American homeland.[25] The statement expanded the scope of the War on Terror to include nations other than Afghanistan. This

---

21. Authorization for 'Use of Military Force,' pub. L. No. 107-40, 2(a), 115 Stat. codified at 50 U.S.C. 1541 (2006)224, 224 (2001))
22. Published, "The 2020 Trump—Taliban "Peace Agreement"—Time to End the War on Terror'" (March 29, 2021)
23. See Afghanistan War Cost, Timeline, and Economic Impact, supra note 4.
24. President George W. Bush, State of the Union Address (29 January 2002) Transcript available at http://georgewbush-whitehouse.archives.gov/news/releases/2002/01/20020129-11.html
25. Gary L Gregg, George W. Bush: Foreign Affairs, UVA Miller Center, https://millercenter.org/president/gwbush/foreign-affairs (The Bush administration's concerns About WMDs in the hands of rogue nation-states) (Last visited 3/16/2022)

policy is reflected in the United States decision to use a U.S-led coalition army to remove Saddam Hussein from power when it was believed he possessed weapons of mass destruction.

"On March 19, 2003, President Bush announced to the nation that the early stages of military operations against Iraq had begun…"[26] Unfortunately, American intelligence was incorrect, no WMDs were discovered, but the presence of al-Qaida and other insurgent groups, to include ISIS, rose in the region. In 2007, President Bush ordered a significant surge in combat troops in an attempt to stabilize Iraq. Many of the actions taken in Iraq directly impacted the Middle East in its entirety, to include Afghanistan. When Bush left office, Iraq was relatively stable, but things in Afghanistan continued to worsen.

---

26. Congress research Service. RS21405 U.S. Periods of War and Dates of Recent Conflicts, (5 June 2020).

5. President Obama Policies and Afghanistan

"As Commander-in-Chief, I will not allow Afghanistan to be used as safe haven for terrorists to attack our nation again. Our forces therefore remain engaged in two narrow but critical missions — training Afghan forces, and supporting counterterrorism operations against the remnants of Al-Qaeda"

— President Barack Obama, October 15, 2015

There were approximately 30,000 American troops in Afghanistan when President Obama took office in January 2009. The War on Terror continued while military forces were also stationed throughout Iraq, but the focus quickly shifted to Afghanistan. On February 17, 2009, President Obama announced he would be sending an additional 17,000 soldiers to Afghanistan while simultaneously committing to reducing the number of troops in Iraq. His plan was simple, overwhelm the Taliban with numbers, increasing the number of troops in Afghanistan from 30,000 to 140,000. He thought this force might convince the Taliban to the negotiating table,[27] as well as deliver some peace to the region. The results were not as he anticipated. The Taliban still refused to negotiate, and the violence continued. According to historian Julie Marks of the History Channel, when the draw down concluded in 2012, over 1,000 Americans were dead or wounded and the Taliban was stronger after the surge than they had been before the surge.[28]

Even in the midst of increasing violence and unable to convince the Taliban to negotiate, Obama accomplished something of great significance; American Special Forces, in an intrepid night raid, killed the al-Qaida leader, Osama bin Laden at his heavily fortified hideout in Abbottabad, Pakistan.[29] This was a major victory and indicated that Al-Qaida was defeated, leaving the Taliban in power.

In October 2011, President Obama ordered the withdrawal of all U.S. forces in Iraq (about 40,000). Months earlier, on June 22, 2011, the president spoke about the mission in Afghanistan in an address to the nation. The following is an excerpt from his speech:

> "I ordered an additional 30,000 American troops into Afghanistan. When I announced this surge at West Point, we set clear objectives: to refocus on al Qaeda, to reverse the Taliban's momentum, and train Afghan security forces to defend their own country. I also made it clear

---

27. Operation Enduring Freedom Fast Facts, CNN (4 October 2020), https://www.cnn.com/2013/10/28/world/operation-enduring-freedom-fast-facts/.
28. Julie Marks, How Seal Team Six Took out Osama bin Laden, History (2 August 2019), https://www.history.com/news/osama-bin-laden-death-seal-team-six.
29. Julie Marks, How Seal Team Six Took out Osama bin Laden, History (2 August 2019), https://www.history.com/news/osama-bin-laden-death-seal-team-six.

that our commitment would not be open-ended, and that we would begin to draw down our forces this July.

Tonight, I can tell you that we are fulfilling that commitment. Thanks to our extraordinary men and women in uniform, our civilian personnel, and our many coalition partners, we are meeting our goals. As a result, starting next month, we will be able to remove 10,000 of our troops from Afghanistan by the end of this year, and we will bring home a total of 33,000 troops by next summer, fully recovering the surge I announced at West Point. After this initial reduction, our troops will continue coming home at a steady pace as Afghan security forces move into the lead. Our mission will change from combat to support. By 2014, this process of transition will be complete, and the Afghan people will be responsible for their own security.

We're starting this drawdown from a position of strength. Al Qaeda is under more pressure than at any time since 9/11. Together with the Pakistanis, we have taken out more than half of al Qaeda's leadership. And thanks to our intelligence professionals and Special Forces, we killed Osama bin Laden, the only leader that al Qaeda had ever known. This was a victory for all who have served since 9/11. One soldier summed it up well. "The message," he said, "is we don't forget. You will be held accountable, no matter how long it takes."

The information that we recovered from bin Laden's compound shows al Qaeda under enormous strain. Bin Laden expressed concern that al Qaeda had been unable to effectively replace senior terrorists that had been killed, and that al Qaeda has failed in its effort to portray America as a nation at war with Islam—thereby draining more widespread support. Al Qaeda remains dangerous, and we must be vigilant against attacks. But we have put al Qaeda on a path to defeat, and we will not relent until the job is done.

In Afghanistan, we've inflicted serious losses on the Taliban and taken a number of its strongholds. Along with our surge, our allies also increased their commitments, which helped stabilize more of the country. Afghan security forces have grown by over 100,000 troops, and in some provinces and municipalities, we've already begun to transition responsibility for security to the Afghan people. In the face of violence and intimidation, Afghans are fighting and dying for their country, establishing local police forces, opening markets and schools, creating new opportunities for women and girls, and trying to turn the page on decades of war.

Of course, huge challenges remain. This is the beginning—but not the end—of our effort to wind down this war. We'll have to do the hard work of keeping the gains that we've made, while we draw down our forces and transition responsibility for security to the Afghan government.[30]

His speech announced plans to withdraw all troops while continuing to build up and train Afghan forces to eventually assume the role of the majority security force in the region. The question remained whether the staged withdrawal of troops from Afghanistan would result in a vacuum similar to that which happened in Iraq where "the Islamic State of Iraq and al-Sham caused the war on terror to expand exponentially."[31] Initially, Obama's plan for the reduction in forces in Afghanistan proceeded as planned. Ultimately, Obama stuck to his word and withdrew thousands of American soldiers, but thousands remained at the end of his term in office, and he announced an end to combat operations in 2014,

---

30. The White House, Office of the Press Secretary, Press Release of President Obama's address to the nation, June 22, 2011. https://obamawhitehouse.archives.gov/the-press-office/2011/06/22/remarks-president-way-forward-Afghanistan.

31. Lt. Colonel, Jeffrey F. Addicott (U.S. Army Retired) from essay: The 2020 Taliban "Peace Agreement"—Time to End the War on Terror." Nebraska law review. https://lawreview.unl.edu/2020-trump-taliban-peace-agreement%E2%80%94time-end-war-terror. Date accessed 15 March 2022.

leaving around 10,000 troops to train and advise Afghan forces. He vowed to remove the remaining forces by the time he left the Oval Office in 2017, but in a June letter published in *National Interest*, the AP notes, ambassadors to and commanders in Afghanistan urged President Obama to keep troops at the current level of 9,800.³²

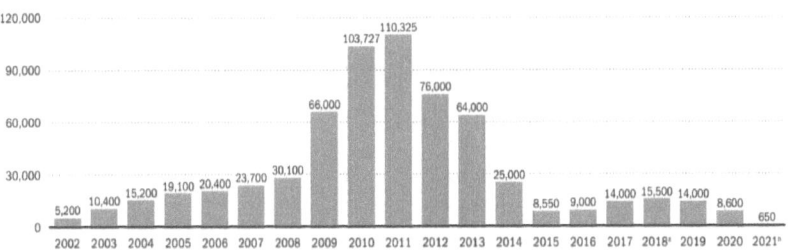

33

---

32. NPR, CHART: How The U.S. Troop Levels In Afghanistan Have Changed Under Obama, July 6, 2016 4:15 PM ET, Danielle Kurtzleben.
33. *Source: Brookings Institution, White House, Credit: Danielle Kurtzleben/NPR*

6. President Trump Policies and Afghanistan

When President Trump entered the Oval Office in January 2017, with a campaign promise to end the War on Terror, which now existed on two fronts, ISIS in Iraq and Syria, and the Taliban in Afghanistan, troop levels had dropped to approximately 14,000. He decided to focus on ISIS first. President Trump stated that we must obliterate ISIS geographically, as he addressed members of the Pentagon to include senior military and national security advisors. Earlier, during his campaign to become president, Trump similarly vowed to "bomb the hell" out of the Islamic State,[34] and declared that the only solution to the problem is a military solution. Jumping ahead, after many years and a series of triumphs and some setbacks, Trump authorized a

---

34. Pamela Engel, "Donald Trump: 'I Would Bomb the s— Out of' ISIS," Business Insider, November 13, 2015, http://www.businessinsider.com/donald-trumpbomb-isis-2015-11

military raid that succeeded in killing the ISIS leader, al-Baghdadi in 2019. Baghdadi's death symbolically marked the destruction of ISIS and permitted Trump to look forward.

Following the great military successes against ISIS in both the Syrian and Iraqi theaters, President Trump turned his attention back to Afghanistan. In his address to the nation in August of 2017, Trump declared that while his "original instinct was to pull out" of Afghanistan altogether, after discussions with senior officials and military leaders, ultimately, he decided to order an increase in American forces in Afghanistan. Trump unveiled a three-pronged approach to his strategy in his address to the nation in August 2017. What appears below is an excerpt from his speech that highlights Trump's three fundamental conclusions:

> "But we must also acknowledge the reality I am here to talk about tonight: that nearly 16 years after September 11th attacks, after the extraordinary sacrifice of blood and treasure, the American people are weary of war without victory. Nowhere is this more evident than with the war in Afghanistan, the longest war in American history—17 years.
>
> "I share the American people's frustration. I also share their frustration over a foreign policy that has spent too much time, energy, money, and most importantly lives, trying to rebuild countries in our own image, instead of pursuing our security interests above all other considerations.
>
> "That is why, shortly after my inauguration, I directed Secretary of Defense Mattis and my national security team to undertake a comprehensive review of all strategic options in Afghanistan and South Asia.
>
> "My original instinct was to pull out—and, historically, I like following my instincts. But all my life I've heard that decisions are much different when you sit behind the desk in the Oval Office; in other words, when you're President of the United States. So, I studied Afghanistan in great detail and from every conceivable angle. After many

meetings over many months, we held our final meeting last Friday at Camp David, with my Cabinet and generals, to complete our strategy. I arrived at three fundamental conclusions about America's core interests in Afghanistan.

"First, our nation must seek an honorable and enduring outcome worthy of the tremendous sacrifices that have been made, especially the sacrifices of lives. The men and women who serve our nation in combat deserve a plan for victory. They deserve the tools they need, and the trust they have earned, to fight and to win.

"Second, the consequences of a rapid exit are both predictable and unacceptable. 9/11, the worst terrorist attack in our history, was planned and directed from Afghanistan because that country was ruled by a government that gave comfort and shelter to terrorists. A hasty withdrawal would create a vacuum that terrorists, including ISIS and al Qaeda, would instantly fill, just as happened before September 11th.

"And, as we know, in 2011, America hastily and mistakenly withdrew from Iraq. As a result, our hard-won gains slipped back into the hands of terrorist enemies. Our soldiers watched as cities they had fought for, and bled to liberate, and won, were occupied by a terrorist group called ISIS. The vacuum we created by leaving too soon gave safe haven for ISIS to spread, to grow, recruit, and launch attacks. We cannot repeat in Afghanistan the mistake our leaders made in Iraq.

"Third and finally, I concluded that the security threats we face in Afghanistan and the broader region are immense. Today, 20 U.S.-designated foreign terrorist organizations are active in Afghanistan and Pakistan—the highest concentration in any region anywhere in the world.[35]"

---

35. Remarks by President Trump on the Strategy in Afghanistan and South Asia, Issued on: August 21, 2017 https://trumpwhitehouse.archives.gov/briefings-statements/remarks-president-trump-strategy-afghanistan-south-asia/

In the remainder of President Trump's speech, he goes on to admonish Pakistan for being "a safe haven for terrorists;" he reveals that he lifted restrictions that limit military leaders from "fully and swiftly waging battle against the enemy;" and he declares that he will "reach out to NATO allies and global partners to support our new strategy with additional troop and funding increases in line with our own" among other significant issues. By February 2020, approximately 13,000 U.S. forces hunkered down in Afghanistan both operating security missions and conducting training for Afghan military and police.[36] By the last months of Trump's presidency, troop levels in Iraq and Afghanistan were reduced to a mere 2,500 soldiers.

On February 28, 2020, Trump was able to get the Taliban and Afghan government to sign a peace deal based on several conditions. In a New York Times article by Mujib Mashal dated February 29, 2020, Mashal quoted president Trump stating, "I'll be meeting personally with Taliban leaders in the not-so-distant future, and will be very much hoping that they will be doing what they say [to include]...killing terrorists...killing some very bad people...[and] keep that fight going.[37]" The deal set the stage for ending the longest war in American history that began following the September 11, 2001 attacks, and ultimately, resulting in the deaths of tens of thousands of people.

---

36. Lolita Baldor &Robert Burns, Pentagon says US Has Dropped to 2,500 Troops in Afghanistan. AP (15 January 2021), https://apnews.com/article/joe-biden-donald-trump-afghanistan-taliban-united-states-16cc1dd5b2f74d463311d212ad0d215a; Idrees Ali, U.S.troops in Afghanistan Now Down to 2,500, Lowest Since 2001: Pentagon, REUTERS (15 January 2021), https://www.reyters.com/article/us-usa-afghanistan-military/u-s-troops-in-afghanistan-now-downto-2500-lowest-since-2001-pentagon-idUSKBN29K229; Courtney Kube & Saphora Smith, Pentagon to Cut Troop Levels to 2,500 in Afghanistan, Iraq, NBC News (17 November 2020), https://www.nbcnews.com/news/military/pentagon-cut-troop-levels-2-500=afghanistan-iraq-n1247992.
37. Mujib Mashal, *Taliban and the U.S. Strike Deal to Withdraw American Troops From Afghanistan*, New York Times, 29 February 2020, et al.

7. President Biden Policies & Afghanistan

When President Biden assumed the Office of the President on January 20, 2021, he inherited a peace deal crafted by Donald Trump. Monikered the DOHA Agreement, the document required the United States, allies, and coalition forces to withdraw all military forces by May 1, 2021. Faced with a concrete exit date, the quickly approaching date of May 1, 2021, and a growing national as well as international pressure to end the war, President Biden faced a very difficult decision. According to Washington Post columnist David Ignatius, "Biden's military advisors had presented him with three unpleasant alternatives [to include]":[38]

1. Leave on May 1 as previously agreed, even though this would probably mean the fall of the Kabul government and a return to civil war.

---

38. David Ignatius, Washinton post. https://www.washingtonpost.com/opinions/2021/12/21/bidens-foreign-policy-tough-guy-vs-pussycat

2. Stay for a limited period, perhaps negotiated with the Taliban, which would delay its eventual takeover.
3. Stay for an undefined period, which could mean a long continuation of what is already the United States longest war.

In an address to the Congress on April 15, 2021, Senator (D) Jack Reed stated, "President Biden has decided to withdrawal from Afghanistan by September 11, 2021 ..." and went on to assert that the decision was one of the toughest President Biden would ever have to make.[39] President Biden made his decision and refused to falter from the set date of departure. Some argue Biden's decision to not address the growing list of concerns as the mass exodus proceeded in Afghanistan as a major lapse in judgement. Conceding that the decisions Biden had to make were immensely difficult, there is merit in that argument. Strategic bases were abandoned, billions of dollars' worth of equipment were left for the terrorists to claim as their own, and much more importantly, America deserted fellow Americans and loyal Afghani advisors and allies.

---

39. Congressional Record S1962, dated April 12, 2021. Accessed 27 February 2022. https://www.congress.gov/congressional-record/volume-167/senate-section/page/S1962?q=%7B%22search%22%3A%5B%22%22%5D%7D

Since the withdrawal from Afghanistan, reports claim as much as eighty billion dollars' worth of U.S. equipment was abandoned in the frenzied evacuation. At this point, there has not been a clear accounting of what equipment and the quantities of each that were left, but the $80,000,000,000 is inflated according to a study completed by the highly touted fact checking organization, POLITIFACT. In an article dated, 1 September 2021, by contributing writer, Tom Kertscher, while addressing a viral image that claimed, "The U.S. left the Taliban an "arsenal" in Afghanistan, including 358,530 assault rifles and 22,174 Humvees," Kertscher found the claim to be mostly false. After scrutinization, it was determined that, "Only a fraction of that amount was spent over the 20 years for hardware; there has been no accounting of how much of it was left and is usable; and one expert told us the aircraft and other military equipment are likely worth less than $10 billion.

Quoting Jonathan Schroden, a military operations analyst at CNA, a safety and security think tank, "…all of the numbers are wrong." The figures included in the viral image were based off 2017, outdated information. In the end, POLITIFACTS study concluded with the following determination: "Only a fraction of that amount was spent over the 20 years for hardware; there has been no accounting of how much of it was left and is usable; and one expert told us the aircraft and other military equipment are likely worth less than $10 billion."[40]

Regardless, any equipment left, no matter the value, several million to billions of dollars, is too much to be left for known fanatical terrorists. Not only does the equipment provide additional armament, the acquisition of U.S. military equipment places top

---

40. Tom Kertchser, POLITIFACT, "Claim overstates military weapons, equipment US left to Taliban in Afghanistan." 1 September 2021. https://www.politifact.com/factchecks/2021/sep/01/viral-image/claim-overstates-military-weapons-equipment-us-lef/

secret technological information at risk as well as provides the enemy with a propaganda boost.

Even as military forces were evacuating, the United States had plans to send additional aircraft to Afghan Forces in July, including several Blackhawk helicopters and A-29 aircraft, while fully realizing that the equipment could very well fall into enemy hands. President Biden was faced with another mammoth decision and took the risk that the equipment could fall into the hands of the Taliban.[41] In the end, the symbolic date to withdraw on the anniversary of 9/11 proved to be at a minimum, a questionable and resultantly, tragic choice.

---

41. Rebecca Kheel. "Billions of U.S. weaponry seized by Taliban." 8/19/2021. The Hill. https://thehill.com/policy/defense/568493-billions-in-us-weaponry-seized-by-taliban

8. President Trump's Deal

On August 21, 2017, while at Fort Myer, Virginia, in a presidential address to the nation, President Trump declared:

> [T]he American people are weary of war without victory. Nowhere is this more evident than with the war in Afghanistan, the longest war in American history—17 years. I share the American people's frustration. I also share their frustration over a foreign policy that has spent too much time, energy, money, and most importantly lives, trying to rebuild countries in our own image, instead of pursuing our security interests above all other considerations.[42]

Trump entered the oval office in January 2017, after campaigning on the promise to end the War on Terror. The war continued despite the best efforts of Trump's predecessors to stabilize the region to a point that a withdrawal was possible. The war was dragging on in Afghanistan when Trump was inaugurated and, just like President Obama, he inherited the war. Satisfied that ISIS was going down in defeat, President Trump turned his attention

---

42. President Donald J. Trump, Remarks on the Strategy in Afghanistan and South Asia (Aug. 21, 2017) [hereinafter President Donald J. Trump Remarks] (transcript available at https://china.usembassy-china.org.cn/remarks-president-trump-strategy-afghanistan-south-asia/).

to Afghanistan, and in late August 2017, he admitted that while his "original instinct was to pull out" of Afghanistan completely, he decided instead to order an unspecified increase in U.S. troop presence to signal to the Taliban that there were no timelines that would drive the United States out of the country.[43]

By 2020, troop levels were reduced to approximately 2,500 American military personnel, the lowest number since the onset of the War on Terror.[44] Unlike his predecessors, President Trump's "strategy [of] a mixture of firm resolve, persuasion, that ever-unpredictable element of good timing," eventually brought the Taliban to the negotiating table.[45] The peace accord, referred to as the DOHA Agreement, consisted of the following (Next three pages consists of the document in its unclassified entirety). An abbreviated overview follows:[46]

---

43. Philip Elliott, *Trump Tries Presidential, Before Reverting to Old Habits*, Time (Aug. 24, 2017), http://time.com/4913683/trump-tries-presidential-reverts-old-habits/ (discussing President Trump's announcement of troop increases in Afghanistan). Originally quoted in Jeffrey F. Eddicott's publication titled: "The 2020 Trump—Taliban "Peace Agreement"- Time to end the War on Terror.'" https://lawreview.unl.edu/2020-trump-taliban-peace-agreement%E2%80%94time-end-war-terror

44. *See* Robert Burns & Lolita C. Baldor, *Pentagon Says US Has Dropped to 2,500 Troops in Afghanistan*, AP (Jan. 15, 2021), https://apnews.com/article/joe-biden-donald-trump-afghanistan-taliban-united-states-16cc1dd5b2f74d463311d212ad0d215a; Idrees Ali, *U.S. Troops in Afghanistan Now Down to 2,500, Lowest Since 2001: Pentagon*, Reuters (Jan. 15, 2021), https://www.reuters.com/article/us-usa-afghanistan-military/u-s-troops-in-afghanistan-now-down-to-2500-lowest-since-2001-pentagon-idUSKBN29K229; Courtney Kube & Saphora Smith, *Pentagon to Cut Troop Levels to 2,500 in Afghanistan, Iraq*, NBC News (Nov. 17, 2020), https://www.nbcnews.com/news/military/pentagon-cut-troop-levels-2-500-afghanistan-iraq-n1247992.

45. Lt. Colonel, Jeffrey F. Addicott (U.S. Army Retired) from essay: The 2020 Taliban "Peace Agreement"—Time to End the War on Terror." Nebraska law review. https://lawreview.unl.edu/2020-trump-taliban-peace-agreement%E2%80%94time-end-war-terror. Date accessed 15 March 2022.

46. Document provided and accessed 3/21/2022: https://www.state.gov/wp-content/uploads/2020/02/02.29.20-US-Afghanistan-Joint-Declaration.pdf

Left Behind in Afghanistan

Joint Declaration between the Islamic Republic of Afghanistan and the United States of America for Bringing Peace to Afghanistan

[February 29, 2020 which corresponds to Rajab 5, 1441 on the Hijri Lunar calendar and Hoot 10, 1398 on the Hijri Solar calendar]

The Islamic Republic of Afghanistan, a member of the United Nations and recognized by the United States and the international community as a sovereign state under international law, and the United States of America are committed to working together to reach a comprehensive and sustainable peace agreement that ends the war in Afghanistan for the benefit of all Afghans and contributes to regional stability and global security. A comprehensive and sustainable peace agreement will include four parts: 1) guarantees to prevent the use of Afghan soil by any international terrorist groups or individuals against the security of the United States and its allies, 2) a timeline for the withdrawal of all U.S. and Coalition forces from Afghanistan, 3) a political settlement resulting from intra-Afghan dialogue and negotiations between the Taliban and an inclusive negotiating team of the Islamic Republic of Afghanistan, and 4) a permanent and comprehensive ceasefire. These four parts are interrelated and interdependent. Pursuit of peace after long years of fighting reflects the goal of all parties who seek a sovereign, unified Afghanistan at peace with itself and its neighbors.

The Islamic Republic of Afghanistan and the United States have partnered closely since 2001 to respond to threats to international peace and security and help the Afghan people chart a secure, democratic, and prosperous future. The two countries are committed to their longstanding relationship and their investments in building the Afghan institutions necessary to establish democratic norms, protect and preserve the unity of the country, and promote social and economic advancements and the rights of citizens. The commitments set out here are made possible by these shared achievements. Afghan and U.S. security forces share a special bond forged during many years of tremendous sacrifice and courage. The Islamic Republic of Afghanistan and the people of Afghanistan reaffirm their support for peace and their willingness to negotiate an end to this war.

The Islamic Republic of Afghanistan welcomes the Reduction in Violence period and takes note of the U.S.-Taliban agreement, an important step toward ending the war. The U.S-Taliban agreement paves the way for intra-Afghan negotiations on a political settlement and a permanent and comprehensive ceasefire. The Islamic Republic of Afghanistan reaffirms its readiness to participate in such negotiations and its readiness to conclude a ceasefire with the Taliban.

The Islamic Republic of Afghanistan furthermore reaffirms its ongoing commitment to prevent any international terrorist groups or individuals, including al-Qa'ida and ISIS-K, from using Afghan soil to threaten the security of the United States, its allies, and other countries. To accelerate the pursuit of peace, the Islamic Republic of Afghanistan confirms its support for the phased withdrawal of U.S. and Coalition forces subject to the Taliban's fulfillment of its commitments under the U.S.-Taliban agreement and any agreement resulting from intra-Afghan negotiations.

The Islamic Republic of Afghanistan and the United States therefore have made the following commitments:

DOHA Agreement: Part One

The Islamic Republic of Afghanistan and the United States recognize that al-Qa'ida, ISIS-K and other international terrorist groups or individuals continue to use Afghan soil to recruit members, raise funds, train adherents, and plan and attempt to conduct attacks that threaten the security of the United States, its allies, and Afghanistan. To address this continuing terrorist threat, the Islamic Republic of Afghanistan and the United States will continue to take the following steps to defeat al-Qa'ida, its affiliates, and other international terrorist groups or individuals:

1. The Islamic Republic of Afghanistan reaffirms its continued commitment not to cooperate with or permit international terrorist groups or individuals to recruit, train, raise funds (including through the production or distribution of narcotics), transit Afghanistan or misuse its internationally recognized travel documents, or conduct other support activities in Afghanistan, and will not host them.

2. The United States re-affirms its commitments regarding support for the Afghan security forces and other government institutions, including through ongoing efforts to enhance the ability of Afghan security forces to deter and respond to internal and external

threats, consistent with its commitments under existing security agreements between the two governments. This commitment includes support to Afghan security forces to prevent al-Qa'ida, ISIS-K, and other international terrorist groups or individuals from using Afghan soil to threaten the United States and its allies.

3. The United States re-affirms its readiness to continue to conduct military operations in Afghanistan with the consent of the Islamic Republic of Afghanistan in order to disrupt and degrade efforts by al-Qa'ida, ISIS-K, and other international terrorist groups or individuals to carry out attacks against the United States or its allies, consistent with its commitments under existing security agreements between the two governments and with the existing understanding that U.S. counterterrorism operations are intended to complement and support Afghan security forces' counterterrorism operations, with full respect for Afghan sovereignty and full regard for the safety and security of the Afghan people and the protection of civilians.

4. The United States commits to facilitate discussions between Afghanistan and Pakistan to work out arrangements to ensure neither country's security is threatened by actions from the territory of the other side.

DOHA Agreement: Part Two

The Islamic Republic of Afghanistan and the United States have consulted extensively on U.S. and Coalition force levels and the military activities required to achieve the foregoing commitments including through support to Afghan security and defense forces.

Subject to the Taliban's fulfillment of its commitments under the U.S.-Taliban agreement, the Islamic Republic of Afghanistan, the United States, and the Coalition jointly assess that the current levels of military forces are no longer necessary to achieve security objectives; since 2014, Afghan security forces have been in the lead for providing security and have increased their effectiveness. As such, the parties commit to take the following measures:

1. The United States will reduce the number of U.S. military forces in Afghanistan to 8,600 and implement other commitments in the U.S.-Taliban agreement within 135 days of the announcement of this joint declaration and the U.S.-Taliban agreement and will work with its allies and the Coalition to reduce proportionally the number of Coalition forces in Afghanistan over an equivalent period, subject to the Taliban's fulfillment of its commitments under the U.S.- Taliban agreement.

2. Consistent with the joint assessment and determination between the United States and the Islamic Republic of Afghanistan, the United States, its allies, and the Coalition will complete the withdrawal of their remaining forces from Afghanistan within 14 months following the announcement of this joint declaration and the U.S.-Taliban agreement, and will withdraw all their forces from remaining bases, subject to the Taliban's fulfillment of its commitments under the U.S.-Taliban agreement.

3. The United States re-affirms its commitment to seek funds on a yearly basis that support the training, equipping, advising, and sustaining of Afghan security forces, so that Afghanistan can independently secure and defend itself against internal and external threats.

4. To create the conditions for reaching a political settlement and achieving a permanent, sustainable ceasefire, the Islamic Republic of Afghanistan will participate in a U.S.-facilitated discussion with Taliban representatives on confidence building measures, to include determining the feasibility of releasing significant numbers of prisoners on both sides. The United States and Islamic Republic of Afghanistan will seek the assistance of the ICRC to support this discussion.

5. With the start of intra-Afghan negotiations, the Islamic Republic of Afghanistan commits to start diplomatic engagement with members of the UN Security Council to remove members of the Taliban from the sanctions list with the aim of achieving this objective by May 29, 2020, and in any case no later than 30 days after finalizing a framework agreement and a permanent and comprehensive ceasefire.

DOHA Agreement: Part Three

1. The United States will request the recognition and endorsement of the UN Security Council for this agreement and related arrangements.

2. The United States and the Islamic Republic of Afghanistan are committed to continue positive relations, including economic cooperation for reconstruction.

3. The United States will refrain from the threat or the use of force against the territorial integrity or political independence of Afghanistan or intervening in its domestic affairs.

4. The United States will continue to work to build regional and international consensus to support the ongoing effort to achieve a political settlement to the principal conflict in Afghanistan.

* * *

After eighteen months of negotiations, Trump struck a deal with the Taliban but had to make many concessions to bring the peace agreement to fruition. A major aspect of the agreement established that the United States would agree to withdraw all United States military forces, its allies, and coalition forces fourteen months from the date of the agreement, thus officially ending the War on Terror by providing a concrete date of withdrawal. Building up to and following a "phased withdrawal" of military personnel, the United States agreed to a prisoner exchange where the U.S. would release 5,000 Islamic Emirate of Afghanistan prisoners and the Taliban would release approximately 1,000 prisoners by March 20, 2020. The United States additionally agreed to review all sanctions (with a goal of removing them by August 27, 2020). This concession was conditional and not an absolute.

The Taliban on the other hand, agreed to cease armed attacks, pledged to shun terror groups such as ISIS and al-Qaida, and to coexist with the current government in Afghanistan. The Taliban also pledged to prevent the use of Afghanistan soil by other terrorist groups for recruiting, training, and fundraising activities.

There are many more aspects of the final agreement as listed in the original document above, but some "classified" portions of the peace accord may never be known by the American public.

Many critics criticize Trump's agreement with the Taliban, declaring the specifics of the agreement were unbalanced, citing that America makes several concrete concessions while the Taliban agree to a series of abstract demands that ultimately, depended on a proven dishonest group of terrorists to suddenly become trustworthy and honorable. In the end, he was the first president in history to successfully negotiate a deal with the Taliban and if the withdrawal were properly managed as he had planned, there is a near-certain chance that Americans and Afghan allies would not have been abandoned.

9. "No Man Left Behind" Policy & Afghanistan

"No Man Left Behind," the longstanding undocumented policy of the United States, is more specifically an assumed mandate of the United States Military. While all branches have their differences, the belief that No Man should be Left Behind resounds across all military organizations. Much like the profiles of heroes presented earlier in this book, there are thousands of stories around that exemplify this precept. From World War I to the War in Afghanistan and every war in between, soldiers have put their life on the line for their fellow Americans and foreign allies.

In fact, the United States Military's "No Man Left Behind" policy can be traced all the way back to the French and Indian War and beyond. Roger's Raiders lived by a set of well-regimented rules and one such policy was to make sure every member of his Ranger unit be accounted for. Roger's set of regulations established a standard that has been carried forward throughout generations all the way to our modern military.

Since the chaotic withdrawal of American forces from Afghanistan, the press and many independent journalists compared the hasty pullout to the fall of Saigon. While many similarities exist between the two evacuations, including the pandemonium caused by a hostile force closing in on the capitol, to the massive evacuation of thousands of people, the comparison between the two wars will continue to be discussed and interpreted by historians for decades to come.

When President Ford made what must have been a gut-wrenching decision to order the evacuation from Vietnam, the North Vietnamese were closing in on the capitol and had already captured many major South Vietnamese cities, to include the second largest city of Da Nang. President Ford had a huge disadvantage as compared to his modern counterparts. In all other American involved wars, no such mass evacuation ever occurred. No precedent existed to inform President Ford as to how to manage a safe and orderly evacuation. Additionally, a peace accord had not been established prior to the Vietcong attack.

On the other hand, President Biden and his administration had the advantage of history, a yardstick by which to guide them as they planned and monitored the evacuation. This was America's chance to avoid making the same mistakes made in Vietnam, but instead, a series of ill-conceived decisions ended in a humanitarian crisis. The mistakes of the past informed very few if anyone in the Biden administration, leading to a series of monumental mistakes.

Individuals awaiting evacuation aboard an Army C130 transport plane.

Instead, in the wake of a flawed evacuation, an evacuation that President Biden claimed to be an extraordinary success[47], President Biden went on the attack. In an address televised on C-Span, President Biden angrily blamed Trump for the debacle in Afghanistan. Biden claimed, "that he had no choice but to end things the way he did in Afghanistan. No choice! No! Choice!"[48]

He couldn't have been more wrong. President Biden systematically overturned one Trump policy after another when he came into office; he could have just as easily decided not to honor Trump's deal with the Taliban. In fact, he did disregard the deal when he decided to delay the May 1st withdrawal that the DOHA Agreement called for. Biden gave the Pentagon until 31 August 2021 to complete

---

47. Reuters. "U.S. President Biden calls evacuation from Afghanistan an extraordinary success." 31 August 2021. https://www.reuters.com/world/us-president-biden-calls-evacuation-afghanistan-an-extraordinary-success-2021-08-31/
48. Podhoretz, John. "An Angry Biden blames Trump for Afghan pullout—then takes credit for it." New York Post, https:/New York Post.com/202108/31//Biden-blames-Trump-for-Afghan-pullout-then-takes-credit-for-it /

the withdrawal of the remaining troops in Afghanistan, and later dispatched approximately three thousand additional troops to the airport to assist with the withdrawal.[49]

Therefore, while he claims he "had no choice," he had the authority to delay the withdrawal until all Americans and Afghan allies could be safely evacuated. Instead, an uncertain number of Americans, and Afghan allies are trapped behind enemy lines. A strong leader does not blame but assumes responsibility for their actions; the buck stops with the leadership.

Even as the Taliban conquered one Afghan city after another, the American withdrawal continued. As they converged on the capital of Kabul and the Airport, many of those promised to be taken care of were left behind in the carnage.

On August 20, 2021, in the East Room of the White House, President Biden stated, "We're going to do everything—everything that we can to provide safe evacuation for our Afghan allies, partners, and Afghans who might be targeted if—because of their association with the United States."[50] Afghan allies served in many roles during the 20-year war, to include service as guides, military attaches, interpreters, and Afghan civilians who risked their lives to bring peace and a level of equality to their people. Instead of paving the way for a safe path to evacuation, Afghans and American's alike were left to fend for themselves.

In the same speech, President Biden went on to state, "The evacuation is dangerous [and] involves risks to our armed forces, and it is being conducted under difficult circumstances." The dangerous circumstances American forces faced, the imminent

---

49. Burns, Robert & Ellen Knickmeyer. Associated Press, 13 August 2021. "Rush of troops to Kabul tests Biden's withdrawal deadline."
50. Dals, Doug. "No man left behind." Freethink.com. 25 May 2020.

dangers that Afghan Allies continue to experience today, were a result of a poorly planned and mismanaged evacuation.

In an MSNBC column authored by Zeeshan Aleem, dated August 31, 2021, he questioned, "was there a significantly better way to withdraw from Afghanistan?" To answer this question, Aleem looked to Anand Gopal[51] a trained sociologist and a renowned foreign affairs reporter who lived in Afghanistan for years. He also embedded with the Taliban and speaks the local languages of the war-torn area.

In the interview, Gopal suggests the better way to withdrawal from Afghanistan was to begin a transfer of power to the Taliban while the U.S. was still in the country. He goes on to assert that Afghanistan was an American construct that had no real legitimacy and that the real war is between not the United States (with Afghan Army) and the Taliban but clarifies that the war is an ongoing "civil war" between different sections of society.[52] In the end, it doesn't take an expert to determine that a military presence was necessary to foster a successful evacuation from Afghanistan.

While Gopal's assertions hold considerable merit, wouldn't a peace accord that required the inclusion of Afghan government's input have been a more feasible solution. In the eyes of the U.S., a legitimate Afghan government was critical to bringing a relative peace to the region. The requirement of the DOHA agreement was to foster a "political settlement resulting from intra-Afghan

---

51. An award wining Journalist who reports for the New Yorker and published the acclaimed book, *No Good Men Among the Living: America, the Taliban, and the War Through Afghan Eyes.*
52. Aleen, Zeeshan. MSNBC Opinion Columnist. "Interview with Anand Gopal," 31 August 2021.

dialogue and negotiations between the Taliban and an inclusive negotiating team of the Islamic Republic of Afghanistan.[53]"

Representatives of the Afghan government (The Islamic Republic of Afghanistan) went on to welcome the proposed reduction in violence period while the U.S. paved the way for intra-Afghan negotiations on a political settlement and a permanent and comprehensive ceasefire. However, the United States did not pave any road to negotiation between the Afghan government and the Taliban. With the vacuum left by the rapid withdrawal of American and coalition forces, there was no opportunity to construct an avenue to any form of mediation.

Regardless the fact that the chaotic withdrawal inhibited any hopes of establishing a bridge between Afghan government and Taliban officials, few, if any, Americans wanted to extend the war in Afghanistan. Casualties of any type are unacceptable but then again, are sometimes a necessity to ensure peace and or at a minimum, create the possibility of harmony among differing cultures. In a Casualty Status Report released by the Department of Defense, the following numbers were released in relation to Operation Enduring Freedom in Afghanistan:

---

53. The Islamic Republic of Afghanistan is a member of the United Nations and is recognized by the United States and the international community as a sovereign state under international law.

OEF U.S. Military Casualties	Total deaths	K.I.A.	Non-Hostile	Pending	WIA
Afghanistan Only[54]	2,218	1,833	385	1	20,093
Other Locations[55]	130	12	118	0	56
O.E.F. U.S. DOD Civilian Casualties[56] [57]	4	2	2	0	
Worldwide Total	2,352	1,847	505	1	20,149

Over twenty-one hundred American soldiers sacrificed their lives in an effort to bring peace to Afghanistan. Each soldier selflessly dedicated their life, heroically fought to better the world, and what is their legacy following the erasure of all progress (that they created with blood, sweat, and tears), that was made towards building democracy in Afghanistan with the seemingly uncoordinated and ill-planned withdrawal?

While every one of their lives was irreplaceable, the tragic loss unacceptable, the very progress they fought to maintain was eliminated in a matter of months. The people they and others fought to protect were left vulnerable and while many were evacuated, an undetermined number of Americans, Afghan loyal

---

54. OPERATION ENDURING FREEDOM (Afghanistan only) includes casualties that occurred between Oct. 7, 2001, and Dec. 31, 2014, in Afghanistan only.

55. OPERATION ENDURING FREEDOM (other locations) includes casualties that occurred between Oct. 7, 2001, and Dec. 31, 2014, in Guantanamo Bay (Cuba), Djibouti, Eritrea, Ethiopia, Jordan, Kenya, Kyrgyzstan, Pakistan, Philippines, Seychelles, Sudan, Tajikistan, Turkey, Uzbekistan and Yemen. Wounded in action cases in this category include those without a casualty country listed.

56. Operation Enduring Freedom, the official name used by the U.S. Government for the Global War on Terrorism.

57. U.S. Department of Defense. "Immediate Release Casualty Status as of 10 a.m. EDT April 4, 2022." https://www.defense.gov/casualty.pdf

interpreters, women activists, and advisors were abandoned, left to defend themselves in the heart of enemy territory. Even as President Biden declared, "The extraordinary success of this mission [Operation Enduring Freedom]..." Americans, Afghan allies, and female activists were fighting for survival. Many being hunted and murdered.

About the only thing President Biden did get correct during his speech of August 31, 2021, in the State Dining Room was to recognize the "...incredible skill, bravery, and selfless courage of the United States Military and...diplomats and intelligence professionals" whose efforts resulted in a true opportunity for peace in the region if managed properly.

So, who exactly, and how many, were abandoned in Afghanistan? Reports vary and a true accountability is currently impossible to compile. In President Biden's address from the State Dining Room on August 31, 2021, he believed there to be 100-200 Americans with some intention to leave. A Senate report released months later (3 February 2022) and compiled by Senator Jim Risch, the ranking member of the Foreign Relations Committee, places the figure at more around 9,000. The report reveals that State Department officials believed that between 10,000 and 15,000 Americans were in Afghanistan as late as Aug. 17.

In the next two weeks, only 6,000 Americans were able to escape the country ahead of the Taliban takeover, leaving anywhere from [4,000 to] 9,000 Americans.[58] Any number more than one is too many, but who else was so callously deserted?

In a recent NBC News report titled, "U.S. Official: 'Majority' of Afghan allies who applied for special visas left behind in Afghanistan,"

---

58. Ginsberg, Michael. "US Government Left As Many As 9,000 American Citizens In Afghanistan After Withdrawal, Senate Report Reveals." *The Daily Caller.* https://Dailycaller.com/2022/02/03/

the Pentagon said about 20,000 Afghans were evacuated to eight U.S. military bases and another 40,000 to bases in the Middle East and Europe. In the same report, a senior State Department official said Wednesday that it appeared a "majority" of Afghans who had worked for the U.S. military and applied for (SIVs)[59] had not been successfully evacuated and remained in Afghanistan.[60] The question becomes why wasn't the Special Immigrant Visa program expedited and the many restrictive hurdles temporarily eliminated to avoid the mass abandonment of American allies and patriots?

The figures in the chart below are based on the Department of Defense annual report of Afghan employment and were analyzed by the Association of Wartime Allies and researchers at American University and estimate the number of Afghan allies left behind. The same group determined that the number could be far higher, depending on assumptions. The group additionally claimed that there may be more than a million Afghans who could be eligible for expedited immigration status (estimated).[61]

---

59. Special Immigrant Visa
60. De Luce, Dan. "U.S. Official: 'Majority' of Afghan allies who applied for special visas left behind in Afghanistan." NBC News, 31 August 2021. U.S. official: 'Majority' of Afghan allies who applied for special visas left behind in Afghanistan (nbcnews.com)
61. Based on Department of Defense employment reports and estimated how many employees filled jobs that were visa eligible. They then estimated the size of their immediate families and subtracted how many have already left the country.

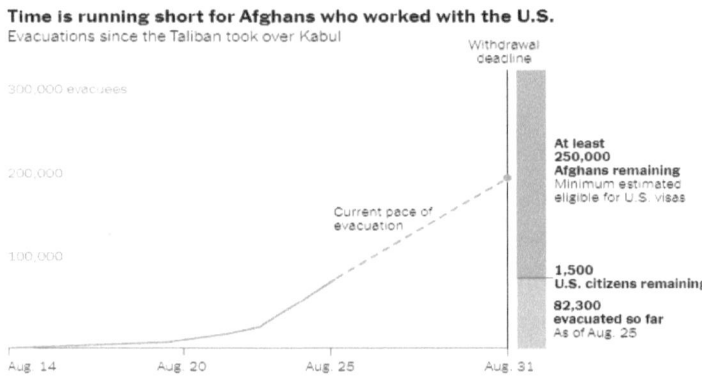

Compiling the exact numbers of Afghan allies left behind becomes extremely convoluted and nearly impossible to determine.

Hundreds of thousands of Afghans qualify for the differing types of visas, green cards, humanitarian parole, those who were in the process of seeking citizenship through other governmental programs, and the many pursuing asylum for a variety of reasons. The many avenues available to citizenship and temporary sanctuary significantly complicates the process of accurately tabulating those who officially qualified for evacuation.

According to an article published in the May 2022 American Legion Magazine, "Tens of thousands of interpreters and their families remained trapped in Afghanistan in the absence of evacuation flights or access to U.S. visas and other documents that would give them a chance of reaching safety.[62]" In any case, the muddled Biden Administration withdrawal resulted in tens of thousands

---

62. Olsen, Ken. "Left behind: Tens of thousands of Afghan interpreters, families are in peril as U.S. bureaucracies quibble over process." *The American Legion: Veteran's Strengthening America," magazine*. (Memorial Day Edition) May 2022.

of Afghan allies trapped behind enemy lines and at risk. No ally should ever be left behind in any situation.

What makes the abandonment of the thousands of Afghan allies even worse is that President Biden himself promised Afghans a refuge. In a speech in July of 2021, President Biden stated, "There is a home for you in the United States if you choose, and we will stand with you just as you stood with us.[63]" In the end, he tragically and inexplicably left thousands of Afghan partners high and dry and vulnerable to the Taliban who hunt them on a daily basis.

---

63. Dale, Daniel. *CNN*. "Fact Check: Biden's dubious claim that 'the law doesn't allow' Afghan translators to be evacuated to US while they wait for visas." 14 July 2021. https://www.cnn.com/2021/07/14/politics/fact-check-biden-law-afghanistan-siv-visas-law-parole-guam/index.html

10. Methods of Qualifying for Evacuation

Chaos ensued as American and coalition forces withdrew from Afghanistan. Thousands of people flooded the airport in panic, so desperate for a way out that several men tried to hold on to a departing plane and fell to their deaths. The country's then president and his cabinet fled as the Taliban rapidly took control of city after city. Only days after the U.S. military withdrew from Afghanistan; the Taliban installed themselves in the presidential palace in Kabul. If the United States knew the date of withdrawal, why was the evacuation process so poorly managed and why wasn't a clear path established for those eligible for evacuation? What was the United States exit strategy? What happened the many months prior to the evacuation date?

Answers to these questions should be sought in order to prevent a similar tragedy from occurring in the future. Regardless, thousands, perhaps hundreds of thousands of eligible and/or people who earned a path to evacuation were left behind.

Senator Robert Portman of Ohio stated, "Too many people were left behind," during an executive session of Congress on November 4, 2021 while speaking in support of those who were left behind in the rush and chaos at Karzai Airport. The questions remain, who exactly was left behind and what was their possible path to salvation?

Unfortunately, most Afghans abandoned in Afghanistan will go unnamed. Most Afghan allies sought asylum utilizing a Special Immigrant Visa. People included in this group are interpreters, drivers working at the United Nations or NATO, Aides, and civilian contract employees. Many already held a Special Immigrant Visa while others had applied for an S.I.V. and were waiting for confirmation.

Members of the House and Congress expected to get a more accurate count of S.I.V. holders/applicants by November 30, 2021. Defense Department reports earlier reported that there were only 700 S.I.V. holders included in the roughly 78,000 people who were [initially] evacuated to the United States. Only 700 of the many thousand S.I.V. holders made it out while the thousands that waited on a response to their S.I.V. applications remained in Afghanistan after the American withdrawal.

Perhaps what is just as alarming as the thousands left to fend for themselves is that an unknown number, the vast majority of evacuees according to ground workers, got past the database screening used by the Department of Homeland Security, the State Department, and intelligence services. Then who exactly was evacuated in the unnecessarily chaotic and unmanaged withdrawal? How many possible terrorists with hopes of further harming America, were ushered into our borders because of the Biden Administrations botched evacuation?

Secretary of Homeland Security, Alejandro Nicholas Mayorkas testified in September 2021, before Homeland Security and the Governmental Affairs Committee stating, "There is a robust screening and vetting process in addition to expertise both in the transit countries and...domestically."[64]

---

64. See Afghanistan (Executive Session); Congressional Record Vol. 167, No. 194 (Senate—November 04, 2021). https://www.congress.gov/congressional-record/2021/11/04/senate-section/article/S7781-1

Senator Portman in a November executive session contradicted Mayorkas's assertion when he revealed that, "An unclassified briefing last week with Federal officials from all of the relevant national security Agencies confirmed what our committee staff had already discovered through our in-person oversight of the vetting operations." The process he mentioned included tours of operations here in the States and also tours of operations overseas at what are called the lily pads, where people are brought from Afghanistan to a foreign destination and then brought to the United States. What they found and what was confirmed at that meeting last week is *"that there is not a robust screening and vetting process for all Afghan evacuees."*

It is true that there is a screening process, which consists of providing fingerprints, a name, and many times, a facial image—so your face, your fingerprints, and your name—to a federal database at the overseas lily pads. But unless the evacuee was a known terrorist, a terrorist affiliate, or a criminal whose name, face, or fingerprints were stored on the system, there *was no vetting*; there were no interviews—nothing else—for Afghans who have been paroled into the United States. Now, here is the problem with that: our database is not complete. Despite repeated attempts to obtain the information, by the way, we don't know how many Afghans were successful in getting past the database screening..."[65]

Portman went on to assert that very few people were picked up or properly added to the database. While most of the Afghan evacuees are peace-loving, good people, a door was left wide open for the entrance of possible terrorists or terrorist sympathizers. If Afghans with no documentation were able to be evacuated in such

---

65. See Afghanistan (Executive Session); Congressional Record Vol. 167, No. 194 (Senate—November 04, 2021). https://www.congress.gov/congressional-record/2021/11/04/senate-section/article/S7781-1

high number, why were so many S.I.V. holders and applicants left behind?

Besides Special Visa Applicants, thousands of at-risk Afghans because of their roles in the government or military, as well as activists or religious minorities, were also left to deal with Taliban reprisal. Despite data indicating that thousands, some estimates reaching over 300,000,[66] President Biden reaffirmed his plan to move all American troops by the end of the month [August], although he did mention the timeline could be adjusted at that time. Many of these people, those not eligible for a Special Immigrant Visa, qualified under other visa programs. Why didn't the administration further delay the withdrawal until all eligible evacuees were safe on American and allied grounds?

The Special Immigrant Visa (S.I.V.), is available to Afghan citizens if they meet the following qualifications: Applicants must be a national of Afghanistan; applicants must have been employed in Afghanistan for a period of at least one year between October 7, 2001, and December 31, 2023: by, or on behalf of, the United States government. As of July 30, 2021, the Special Immigrant Visa program requires applicants to have been employed for a minimum of one year, between October 7, 2001, and December 31, 2023. Applicants must also have experienced or be experiencing an ongoing serious threat as a consequence of their employment, which in this case is very easy to determine; all American employed Afghan citizens were at risk.

For those that do not qualify for an S.I.V., additional avenues exist for Afghan citizens through the U.S. Refugee Admission

---

66. Buchanon, Larry & Lauren Leatherby. NY Times, 25 August 2021. https://www.nytimes.com/interactive/2021/08/25/world/asia/afghanistan-evacuations-estimates.html Number compiled by the International Rescue Committee.

Programs that may have qualified some Afghans for evacuation. Under the program, three levels of visas exist (See chart below).

U.S. Refugee Admissions Program	Available to:
Priority 1 Visa	Refugees identified and referred to in the program by the United Nations High Commissioner for Refugees (UNHCR), a U.S. Embassy, or a designated non-governmental organization (NGO)
Priority 2 Visa	Interpreters who worked for the U.S. military but not long enough to qualify for S.I.V. Afghan employees of a U.S. news organization, and Afghans employed directly by Indiana University, Ball State University, Kansas State University, Save the Children, International Assistance Mission, Reuters, Associated press, etc.
Priority 3 Visa	Reserved for family reunification with immediate relatives already in the United States.

In the case that a person did not qualify for any of the visa programs, eligible Afghans could petition for Humanitarian Parole. Afghans or sponsors who live outside of the United States can request humanitarian parole. United States Citizenship and Immigration Services (USCIS) makes determinations for humanitarian parole, but generally, the beneficiary must show particular vulnerabilities and significant family or other ties to the United States. Unlike a visa, parole does not give the beneficiary immigration status in the United States, and they will still have to apply for an adjustment of status if they plan to remain beyond the duration of their parole.⁶⁷ Some human rights and freedom of expression activists might also qualify for humanitarian parole. For instance, many Afghan women would qualify for possible evacuation for humanitarian reasons, especially those who advocated for women's rights and equality during the American influenced peace. Since the collapse of the Afghan government and the U.S. Military withdrawal, women and girls in Afghanistan are now almost entirely excluded from public life after just three short months of full Taliban rule. Gender equality is nonexistent and human dignity is openly attacked via state-sponsored oppression.⁶⁸

While Afghan citizens may qualify for citizenship and refugee status through a variety of programs, the confusing application process in addition to the chaotic withdrawal of Afghans who managed to get to the Kabul Airport for evacuation, resulted in many not passing the flawed and insufficient vetting process. According to a *Stars and Stripes* article dated September 17, 2021, many passports were destroyed during the hasty retreat from the U.S. Embassy, which resulted in many not being able to leave the country, to include S.I.V. holders and applicants alike.⁶⁹

In the end, many avenues exist for an Afghan ally to seek asylum, whether temporary or permanent, in the United States. Interpreters, drivers, at-risk Afghans, refugees, humanitarian parole, and family seeking reunification all had a path to evacuate the country, so why were thousands left behind? The date of withdrawal was established within the Doha Agreement and provided

---

67. Padilla, Alex, U.S. Senator. "Afghanistan Evacuation and Immigration Resources." https://www.padilla.senate.gov/afghanistan-evacuation-resources/
68. Fagan, Paul & Natalie Gonnella-Platts. "Will the US abandon the women and girls of Afghanistan?" 10 December 2021. https://thehill.com/opinion/international/585228-will-the-us-abandon-the-women-and-girls-of-afghanistan/
69. Lawrence J.P. Stars and Stripes. 17 September 2021. "Afghan translators stranded as passports become casualty of US Embassy capture

plenty of time for the proper documentation to be finalized before the withdrawal. A study by the Office of the Inspector General identified eight long-known flaws in the system. Two of the issues were determined to not "significantly affect" the program. These possible barriers include:
1. The required medical examinations are costly yet readily available.
2. The consideration of how will the S.I.V. program impact local hiring practices within the United States.

The third identified obstacle, said to be outside the department's control by the State OIG involved:
3. The availability of visas (Goes directly back to the accountability issues involved with the process).

The O.I.G. went on to confirm that five obstacles identified by Congress, if unaddressed, will remain impediments to implementing the Afghan SIV program and achieving the goals established by Congress, which include issuing an SIV within a nine-month timeframe (A result of the inefficiency and overly complex bureaucracy). The five remaining obstacles consist of:
4. Staffing issues related to the efficient processing of visa applications.
5. Interagency delays as associated with the communication between the various departments and processes involved in the screening requirements.
6. Document processing bottlenecks and a lack of accountability.
7. The existence of a centralized database to compile data and track individual case progress.

8. An inefficient bureaucratic system as a whole.[70]

Why should the visa application process take nine months? Granted some applicants are the source of their own issues and delay, but the inefficiency of the visa system led to tens of thousands, possibly hundreds of thousands of Afghans being abandoned behind enemy lines. In a July speech, President Biden said he was taking steps to speed up the visa process and went on to state, "There is a home for you [Afghans] in the United States if you so choose, and we [American people] will stand with you just as you stood for us." In the same speech, he also goes on to assert that current law does not permit the movement of Afghan translators to the U.S. while awaiting the visa processing. This claim leads to the question, how can migrants from the southern border be allowed into the country while awaiting a decision on their cases while deserving Afghan allies cannot?

CNN's political fact checking service later determined President Biden's claim to be "dubious" since the Department of Homeland Security has the authority to grant migration "parole" for urgent humanitarian reasons or significant public benefit on a case-by-case basis. Adam Bates, policy counsel at the International Refugee Assistance Program, supported CNN's determination when he stated, there is "nothing" in either the parole law or the law governing the Special Immigrant Visa (S.I.V.) programs for Afghans, "that forbids President Biden from utilizing parole to get S.I.V. applicants and their families into the United States where they can be processed along some path to immigrant status."

---

70. Putz, Catherine. "A Glimpse at What Went Wrong with the Afghan SIV Process Two June 2020 State OIG reports illustrate some of the long known, ever-unresolved issues stymieing the Afghan Special Immigrant Visa process." The Diplomat. 12 September 2021. https://thediplomat.com/2021/09/a-glimpse-at-what-went-wrong-with-the-afghan-siv-process/

Did President Biden simply not understand the resources available to him or did he have some sort of personal reason for not fully utilizing the authority he had to help Afghan allies? Regardless the answer, thousands of Afghans are struggling to survive on a daily basis as they attempt to evade Taliban forces who actively hunt, imprison, and in many cases, execute those who openly assisted American and NATO forces.

The time to evacuate Afghans was before the withdrawal of American Forces. Instead, tens of thousands have been abandoned due to the lack of an organized and well-planned exit plan.

12. Portrait of an Abandoned Afghan: Massih

"It's beyond me how he has the SIV approved, all the hard work is done, and the system totally failed him."

—Roger Cartwright (retired military lawyer working with Massih)

Embassy officials were only following protocol when they destroyed sensitive materials before the Taliban takeover. The problem is that this action detrimentally affected many S.I.V. applicants and even those seeking priority visas at all levels. Many translators, advisors, drivers, and assistants depended on the very documentation needed for immigration that the embassy officials destroyed.

One translator who goes by the name Massih was issued his Special Immigrant Visa but when his Afghan passport went missing from the embassy, most likely destroyed as a precaution, he was very far from achieving his goal of immigrating to the United States. Roger Cartwright, a retired military lawyer who worked with Massih stated, "It's beyond me how he [Massih] has the S.I.V. approved, all the hard work is done, and the system totally failed him."⁷¹ Now that the United States no longer has a presence in Afghanistan, U.S. officials will possible have to provide financial assistance to bordering countries in exchange for opening their borders to people without passports (Lawyers and advocates informed the Stars and Stripes) as well as make an attempt to open negotiations with notoriously anti-American Taliban officials.

Did we get all of the Afghan translators out of Kabul in the recent evacuation debacle? No. In fact, an article in Stripes yesterday revealed that literally hundreds of them were not just left behind, but their passports were destroyed in the hasty retreat from the US Embassy. Which means that though they were approved for an SIV, they can't leave the country, even if the Taliban were to let them leave

Massih believes that not only was his passport destroyed but that the passports of his family may have been as well when he left them at the Embassy on August 4, 2021 to receive their S.I.V.s. Ten

---

71. Higbee, Faye. "Afghan Translators Left Behind." 18 September 2021. https://www.unclesams-misguidedchildren.com/afghan-translators-left-behind/

days later, the Embassy notified him that his visas had been issued and would soon be available for pick up along with his passports. The Embassy fell to the Taliban a day later.[72] An unnamed military officer (Who requested anonymity) who worked with Massih, asserts that the Taliban blocked every attempt to get to the airport for evacuation. This leads to a number of questions. First, who exactly was evacuated and what are the odds that they were properly vetted since the majority of S.I.V. holders were abandoned?

The Trump administration agreed to an initial reduction of US forces from 13,000 to 8,600 troops by July 2020, followed by a complete withdrawal by May 1, 2021, if the Taliban kept its commitments.

President Biden, on April 14, 2021, announced a four-month postponement of former President Trump's deadline, stating that the troops "will be out of Afghanistan" before Americans mark the 20th anniversary of "that heinous attack on September 11th." Washington extended the withdrawal date to provide the Ashraf Ghani-led Afghan government more time to set up to fight against the Taliban. The US also needed a considerable time-period to hand over previous US-controlled military bases and tactical equipment to the Afghan forces.[73]

A key takeaway from then President Trump's statement regarding the withdrawal of U.S. forces was his foresight to add the condition that the withdrawal date is flexible depending on "if the Taliban kept its commitments." President Biden later used this to justify delaying the withdrawal to August 31, 2021, but why wasn't

---

72. According to the Stars and Stripes newspaper.
73. Bose, Joydeep | Edited by Amit Chaturvedi, "Why did Biden choose August 31 as deadline for US withdrawal? Here's a timeline," Hindustan Times, New Delhi. https://www.hindustantimes.com/world-news/why-did-biden-choose-august-31-as-deadline-for-us-withdrawal-here-s-a-timeline-101629861240721.html

the date once again delayed in response to the Taliban's failure to meet their obligations set forth in the Doha Agreement? As the withdrawal continued throughout August, newscasts distinctly provided horrifying images of abuse while Taliban efforts to waylay the evacuation process unfolded before the eyes of the world. As a result of inaction, Afghan allies like Massih and thousands like him, were callously abandoned to fend for themselves.

Copy of a Taliban threat letter sent to an unnamed Afghan

13. Portrait of an Abandoned Afghan: Ahmed

"I cannot stay on one location long, but I also cannot know how long I can keep up like this... eventually I feel I will be found."

—Ahmed (Afghan interpreter)

In the early morning hours at approximately 2:30 a.m. a text message was released stating, "[T]here is a major search operation going on in Kabul tonight...If possible try to stay out of sight, and away from central areas." The message went on to instruct recipients to hide documents, computers, and listed the districts of Kabul as an area where the Taliban were targeting Afghan citizens that night as they [the Taliban] moved from house to house.

One Afghan who received this message along with others who worked for the government, Ahmed (Whose name is fictitious for security reasons), already realized he must stay in hiding. He is one of countless Afghans now hiding in fear trying to survive the Taliban who actively attempt to hunt them down. Hiding and staying on the run is one way to survive for fearful Afghans who tactically move during the night, leap-frogging from home to home in an attempt to remain a step ahead of the Taliban. Many of the encrypted messages evolve from U.S. soldiers and veterans who try to help their Afghan friends and allies avoid the Taliban sweeps of villages and cities looking for those who aided the American war effort.

Many other Afghans who worked for the United States are in the same situation and in hiding.[74] According to phone interviews, these same people live in fear that Washington's interest in helping them evacuate is fading, leaving them to fend for themselves.

Ahmed was unable to escape during the chaotic withdrawal despite putting his life on the line while working for the United States Military and State Department for several years. Now, Ahmed and others like him live in fear, stealthily maneuvering as the Taliban & Islamic State try to capture, imprison, or more often, to maim and murder those that supported the United States government. "I cannot stay on one location long, but I also cannot know how long I can keep up like this...eventually I feel I will be found," Ahmed stated.

As reported by Robbie Gramer, et. al., from the FP News, a former contractor for the U.S. State Department and military, who is also in hiding and spoke on condition of anonymity stated, "Even if you put aside the Taliban, ISIS is very dangerous for us [too]." He went on to reveal that they [the Taliban] announced that if they find anyone who worked for the United States in any way, if they find them, they will kill them. "Every day I hear of three or four other former workers for the U.S. Military or government [who were] kidnapped by unknown gunmen and are not seen again." The Biden Administration's decisions and lack of an organized evacuation directly resulted in the death of many American allies.

---

74. Gramer, Robbie, Jack Detsch, & Amy Mackinnon. "Those Left Behind in Afghanistan." *FP News*. 1 October 2021. https://foreignpolicy.com/2021/10/01/afghanistan-biden-interpreters-special-immigrant-visa-evacuation-state-department/

14. Portrait of an abandoned Afghan Khan

"I heard nothing from anyone [America]... that makes you feel left behind, makes you feel you will be shot and dead every day..."

—Khan (Afghan Interpreter)

Another Afghan who goes by the name Khan was also unable to evacuate. In an episode of NPR's *Up First Sunday*, the host, Rachel Martin interviewed Khan who openly questioned, "Why did I help the U.S. Army? Why was I risking my life being in a job as [an] interpreter?" The following is a partial transcript as well as some key things revealed by this interview from Khan.

Rachel Martin, Host: It's hard to shake the images coming out of Ukraine right now—the bombed-out theater where a thousand Ukrainians, many children, were seeking safety, the 3 million refugees who've had to flee because of Russia's invasion of their country. They are safe for now, out of the shelling. But the survivors of another war in a different part of the world are still desperate to get out.

Martin: Last fall, I started talking to an Afghan man who worked as an interpreter for the U.S. military during the early years of the war. Now with the Taliban in control, he is in hiding and terrified for the safety of his family. We'll bring you snapshots of his life from weeks after the Taliban took over the country and what it's taking for him to survive now. We'll also hear from an American veteran who's been working to try to get his Afghan colleagues to safety before it's too late.

Matt Zeller: You kind of had to sort of laugh at just how awful they made the process of trying to get a lifesaving visa on these applicants.

Martin: There's a concept in Afghanistan known as mehman nawazi. It's loosely translated from Dari to English as hospitality, but it really goes beyond that. It's the idea that you give what you can to make a stranger feel truly at home. And it's not a thing where you feel obliged down the road. There are no strings attached, no expectations. After the U.S. drove the Taliban from power in 2001, Afghans extended the same hospitality to the Americans because most Afghans did feel liberated by the U.S. With the Taliban ousted, the soccer stadium in Kabul would no longer be used to publicly execute people who had committed petty crimes or women accused of infidelity. Afghans could dance, play music. Women could take off their burkas, and men could shave their beards. It was a time of immense hope, and Afghans gave Americans credit. But the war didn't stop. Americans hunted Taliban and al-Qaida and later ISIS, one village at a time. In the early years, Afghans welcomed the troops into their homes with tea and small dishes of golden raisins and pistachios. And those conversations, which were really intelligence-gathering operations for the U.S., were impossible without the help of Afghan interpreters.

Khan: The first time I started my job as interpreter was December 2004.

Martin: OK. And you were working for the U.S. military?

Khan: Yes, I started working directly with the U.S. military.

Martin: This is Khan. We're using that name to protect his identity. He worked side-by-side with U.S. forces from 2004 to 2008. And when he signed up, no one said, hey, sign on the bottom line, and we'll hand you a U.S. visa. But many Afghans did have expectations about America's commitment to their country, mainly because America had laid them out.

Martin: Khan believed part of that responsibility meant taking care of Afghans like him who were risking their lives to help U.S. forces.

Khan: We were lucky we survived. We survived around 12 ambushes. We had different combat missions. Normally, when an interpreter goes to a mission, he doesn't get more information—where we were going and when we were going—we were just called to be ready, that there's a mission.

Zeller: Combat, war is the most intense experience a human can endure.

Martin: I want to introduce you to another voice here. This is Matt Zeller. He's an Army veteran who worked closely with Afghan interpreters when he served in Afghanistan in 2008.

Zeller: It is so powerful that you don't come back the same from it. No one does. It is a completely transformative experience. And the Afghans that we went through that with—you know, they might have started out as complete strangers to us upon arrival, but by the time it was time for our unit to leave, saying goodbye to those people was one of the most heartbreaking moments of my life. It felt like I was abandoning fellow members of my unit.

Martin: As U.S. troops kept cycling in and out of the country, they stayed dependent on the eyes and ears of the Afghan interpreters like Khan. I talked with Khan for the first time in September of last year. He told me he had started getting death threats after his first year with the U.S. military.

Khan: We were threatened directly and through phone calls, through letters. One day, I was just walking to my home. Around 10 meters away from my door, two guys on one motorbike started shooting. I ran away where there were a few shops, where the insurgents couldn't come back there. They thought after shooting I would run to my home, and they could kill me easily. But I was lucky. I turned around, and I ran away, back to the shops, so I survived.

Martin: It eventually became too much. Khan constantly feared for his family's safety. So he fled to India. He became a student and finished an undergraduate and a master's degree. He came back to Afghanistan in 2014, but the same threat was there. The Taliban still wanted retribution against anyone who had worked for the U.S. military. Khan remembered a promise that he had heard over and over, that the U.S. government had his back.

Martin: In 2014, Khan applied for his visa to the U.S. It was called a special immigrant visa, and it was meant to streamline the immigration process for Iraqis and Afghans who helped U.S. forces. Matt Zeller learned about the special immigrant visa through the interpreter he had worked with in Afghanistan, a man named Janis Shinwari. Janis saved Matt's life at one point during his deployment, and they became like brothers. Then in 2009, he called Matt over Skype and said the Taliban had put him on a hit list.

Zeller: The Taliban were actively hunting him because of his service to the United States, specifically because of what he had done to save my life. He goes, there's this newly created visa called the special immigration visa. To get it, you would have to nominate me for it. And I said, sure, not a problem, thinking naively that this might take — I don't know — six months, maybe a year.

Martin: It took four years. The same year Khan applied for the special immigrant visa, Matt and Janis created a nonprofit to help Afghans climb their way through the mountain of paperwork related to SIVs.

Martin: This was a challenge for Khan. He worked for a rotating cast of colonels and captains during the war, and no one had told him that he should get letters of recommendation from his U.S. supervisors or their contact information before they went back to the U.S.

Zeller: The next thing is not only do you need the American who served with you to write you a letter. You need the third-party contractor who physically cut your paycheck to also verify your service. That contract often changed annually, if not multiple times per year, from company to company. This is all on the Afghan in a country in which, you know, there aren't phone lines in most villages and internet in most places.

Martin: Khan waited six years. All his documents were in order. There was nothing left for him to do, and he says all the while, the threats against him continued. Then, in April 2021, President Biden announced that after two decades, America was done. Biden said he would pull all U.S. troops out of Afghanistan by September. Three months later, he made this promise to Afghans who had fought with the Americans.

The interview continues beyond this excerpt and revisits President Biden's message that "Our message to those women and men is clear. There is a home for you in the United States, if you so choose, *and we will stand with you just as you stood with us.*" Khan goes on to address a myriad of issues to include speaking about the frustration of trying to obtain a Special Immigration Visa, the fall of Kabul to the Taliban, his disappointment regarding not being evacuated, and details about his life on the run following the mass exodus of American forces.

For a short time, it appeared that Khan might receive his S.I.V. The State Department set a time for his immigration interview. However, the interview was rescheduled on August 11, 2021 due to Covid protocol. When finally interviewed, Khan expressed that he was from Kandahar and stressed how important it was to get his family out of Afghanistan. By now, many key cities had already fallen to the Taliban. Four days later, the Taliban marched into Kabul and took control of the government. Khan left his home and went to a hotel for a week before becoming frustrated and returning to Kandahar.

At the same time, veteran Matt Zeller who knew of Khan's

case, worked every military and political connection to try to get Afghan interpreters, including Khan, out of the country. He worked around the clock actually operating off Afghan time to make sure he didn't miss any important phone calls. While Zeller waited for some sort of contact, Khan waited for a call from the State Department that never came.

Matt Zellers organization, *No Man left Behind* helped evacuate thousands of Afghans, but the vast majority who went through the process of acquiring a Special Immigrant Visa never got out. Many simply didn't have the necessary paperwork, others were unable to get to the airport due to efforts of the Taliban to complicate the process. When Rachel Martin asked, "So how does that make you feel (being ignored and left behind)" Khan answered, "That makes you feel disappointed. That makes you feel left behind. That makes you feel you will be shot and dead every day you're living in Kandahar, where all the Taliban's around you. That makes you feel you have no hope."

Now, Khan lives day by day in hopes of getting a call from the United States stating his application for a visa is approved, but even if he one day receives that call, it may be too late. The Taliban have no interest in letting those that collaborated with the United States fly off to freedom. When asked what life is like now (since the Taliban takeover), Khan explains how he had to move his family for his and their safety to a place where no one knew him until he received some answers. When asked do you know where to get those answers (regarding the status of his visa), Khan replied, "Yes, I know someone is asking God to help me. That's it. I—that's all I have. There's no one else I can call, I can't email, I can't contact. No one else. No one. I just have to keep praying, see what happens."

So, Khan, an Afghan patriot who selflessly allied with America in order to build a better Afghanistan, lives in fear trying to keep his family safe. Khan, an American ally, allies who President Biden declared, "There is a home for you in the United States…we will stand with you just as you stood with us," has heartbreakingly been abandoned, living in fear for his life, fighting for the survival of his family each day.

God bless all of the brave Afghans' who truly endeavor to build a free and equitable Afghanistan. Especially pray for the thousands left behind, fighting to stay alive because our government reneged on their promise to provide a home for the Afghan patriots who courageously stood side-by-side with the U.S. Military. What kind of message does this send to our allies around the world?

15. Portrait of an Abandoned Afghan: Romal Noori

"We basically acted like a bridge between Americans and the locals... We were there to help them [US military]... Now we need their help."

—Romal Noori (translator)

"[I] fear being targeted if the Taliban reach[es] Bagram," Romal

Noori shared in a New York Times article and later on video. For a good part of his life, Romal worked for the United States, but after nine years of service, working as a janitor, a machine operator, and later, as an interpreter, he lives in [75] fear that the Taliban might single him out for his service to the United States Military. He grew up near Bagram Air Base, once the largest American Military Base in Afghanistan, and eventually began working for the U.S. as a teenager.

Now 31, Romal lives with his wife and four children. Since the U.S. withdrawal, the economy has tanked, making it difficult to survive from day to day. Add in the lack of security and the Taliban closing in, he fears his family may be targeted. In a *New York Times* article dated August 10, 2021, journalist Isabelle Niu interviewed Mr. Noori in a series of video calls. While offered anonymity, Romal decided it best to appear on camera in hopes of "draw[ing] the attention to the plight of Afghan laborers who risked their lives working for the United States and are now counting on the Biden administration to help them leave Afghanistan."

"We basically acted like a bridge between the Americans and the locals," Romal said. He goes on to say that they were there to help but now he finds himself in need of their (the military/U.S. government's) help. He applied for a Special Immigrant Visa (S.I.V.) offered under the umbrella of the Afghan Allies Protection Act of 2009. His visa was delayed and ultimately, never received

---

75. Niu, Isabelle, Danielle Miller, Khyber Khan, & Nicole Fineman. Image compliments of the *New York Times* shows Romal Noori in Afghanistan. "We Are Are Alone' An Afghan Translator's Plea for Help." 10 August 2021. https://www.nytimes.com/video/world/100000007887080/afghan-interpreter-us-military-video.html

since U.S. troops pulled out, and the Afghan government collapsed under the auspices of the Taliban invasion.[76]

After troop evacuation, he no longer had avenues like Facebook, email, or even a phone to reach out. Romal lost touch with his military contacts. He now lives in hopes of the U.S. government informing him that his family's visa was approved, but he is losing faith as each day passes. On July 21, 2021, the State Department announced an evacuation plan titled Operation Allied Refuge with a goal of evacuating SIV applicants in a timely manner. Romal once again had hope but shortly after, he discovered that he was not eligible. To qualify for the program, the Special Immigrant Visa must have been approved. Unfortunately, his S.I.V. never reached the approval process.

Romal Noori tragically is still stranded in Afghanistan, hoping to be evacuated before the Taliban reach his family. He courageously served side by side with the United States military, making friends and contacts through his nine years of service and now he finds himself abandoned when he needs help himself. As his interview neared an end, a noticeably shaken, Romal stated, "Me, like, I'm gone. I don't have no more wishes, you know? The only wish that I have is that I want to see my kids happy. That's all. I'm trying to do the best, but there's no...nothing I can do now for them." Thankfully. After the efforts of many people and organizations, Romal and his family were eventually evacuated. Tragically, thousands of other families will never experience their dream of coming to America.

---

76. Niu Isabelle. "He Worked With the U.S. for 9 Years. Now He's Left Behind in Afghanistan." *New York Times*. 10 August 2021. https://www.nytimes.com/2021/08/10/world/afghan-interpreter-us-military.html

Image above: Judy Lohmeyer greets Romal Noori and his family with open arms as they exit the terminal at the Springfield-Branson National Airport shortly after arriving in Springfield on Monday, Jan. 10, 2022. Romal Noori, who served as a translator for American troops in Afghanistan, fled the country with his family after the troop[77] withdrawal last year.

---

77. Papes, Nathan. "Afghan family meets U.S. veteran friend at airport to start 'new life' in Ozarks." *Springfield News- Leader.* 15 January 2022. https://www.news-leader.com/picture-gallery/news/2022/01/16/afghan-family-romal-noori-meets-us-veteran-tommy-breedlove-airport-start-new-life/6529741001/

16. Portrait of an abandoned Afghan: Zak

"The consequences are, me and my family will be killed"

—Zak (Interpreter)

Zak, a 31-year-old Afghan interpreter worked for Marine Forces in Afghanistan. During a *New York Times* interview, Zak described the battlefield as "horrible [and full of] danger," when asked what his first impressions were when he first began working with the Marines. At first, he thought his efforts—along with the Marine expeditionary forces—might rid Afghanistan of the Taliban, but now he explains it gets worse day by day. Zak's S.I.V. never completed the approval process and now he finds himself trying to survive. He has received threats from the Taliban and according to him, the Taliban is sending the message, "Whenever we find you, we will kill you and your family." The Taliban calls him an infidel since he worked with Americans. Now he feels helpless.

During the phone interview, the line suddenly went dead, and the interviewer began speaking with Major Schueman who was a platoon commander in Q Company, 3rd Battalion, 5th Marines and was deployed to the Helmand Province, Sangin District, Afghanistan from September 2010 through April 2011. Major

Schueman served with Zak and knew him personally. He met Zak early in his deployment and remembers him then as a young man or as he stated, "The same age as many of my Marines, 18, 19. He explains that we 'kind of adopted him into the platoon and protected him.'" He went on to reveal that he was an immediate asset, and he was on patrols with him on almost a daily basis. Zak was at the major's side through enemy contacts, lots of firefights, and ambushes. Zak assisted with questioning of locals and detainees.

Schueman informed the interviewer (Lynsea Garrison), that Zak has a warrior spirit that he exhibited on multiple occasions. Major Schueman recalled one such incident in which his unit was trapped in a minefield while Zak monitored Taliban communications. Zak learned that they were going to initiate an I.E.D.[78] Zak commenced to — without regard for his own life — sprint through the minefield and tackle this guy and then informed the Major that this was the guy talking on the radio about his patrol.

Major Schueman described another incident when he had two Marines blown up, one a triple amputee. His unit was still under fire when Zak picked up the weapon of a wounded Marine and returned fire. Schueman went on to describe Zak as, "Basically one of my Marines [who] became one of our brothers...[and] Zak, I think, felt like he was really a member of his platoon."

When asked, "Did he ever talk about, like, leaving Afghanistan, even in those earlier years? Did he ever talk about that with you?" Major Schueman replied, "Yeah, I think there was a lot of hope that we would reunite on American soil."

---

78. Improvised Explosive Device

Left Behind in Afghanistan

Marine Corps Major Thomas Schueman is determined to get his translator Zak and his family out of Kabul, Afghanistan.[79]

The State Department and President Biden made many promises to Afghan interpreters that never came to fruition. The unnecessarily hasty evacuation left Zak and countless others in the crosshairs of enemy sights. Major Schueman was able to reengage with Zak following President Biden's speech in which he declared, "The United States will begin our final withdrawal."

He asked Zak, "What are the consequences of us leaving?"

Zak replied, "The consequences are, me and my family will be killed."

In the same speech, President Biden went on to state, "We...will not conduct a hasty rush to exit. We'll do it responsibly, deliberately,

---

79. Photo provided by ABC News.

and safely..." But the withdrawal was anything but safe and responsible. Instead, thousands of Afghan allies were deliberately left stranded with no regards for their safety and very little thought of responsibility. Instead, "Exactly 24 hours after the last US military transport plane departed Kabul, leaving hundreds of Americans and thousands of Afghan allies at the mercy of terrorists...President Joe Biden declared the mission an 'extraordinary success.'"[80] In what world could such a flawed, unorganized, and chaotic evacuation ever be called anything but a massive failure?

In an ABC News interview, Major Scheuman declared that "He [Zak] wasn't just a translator, he was my brother, basically one of my Marines...I think it's a very simple transaction. You serve with U.S. Forces, and we will provide you a visa...he served with U.S. Forces, we did not provide the visa. I think that's a betrayal."[81] Schueman later stated, "I have a lifelong commitment to the people I serve and lead." When able to reach Zak in a later phone call, Zak's babies (he has four children) were heard crying in the background. His family is packed and waiting on word to roll out.

After two failed attempts, Zak and his family were met by Schueman who stated, "We made a third attempt at the airport with Zak and his family...there was a lot of confusion and a lot of chaos." But Major Schueman arranged for a point of contact on the ground who linked up with Zak at the gate. He and his family were pulled in and he was manifested for a flight.

---

80. Editorial Board. With Americans, allies left behind, a clueless Biden declares Afghan mission 'extraordinary success," *The New York Post*. https://nypost.com/2021/08/31/a-clueless-biden-declares-afghan-mission-extraordinary-success/

81. Kim, Deborah, Allie Yang, Ashley Louszko, & Jake Lefferman. "Victory after Marine's tireless battle to rescue translator from Kabul." *ABC News*. https://abcnews.go.com/International/marines-uphill-battle-rescue-afghan-translator-kabul/story?id=79501057

Zak's last recorded statements provided great hope of his eventual evacuation. Zak declared, "We've just done our process, and waiting for our flight in the terminal…We are set now, and let our guys know [I assume the Marines of the Q Company, 3rd Battalion] Zak is set and waiting for his flight."

Major Schueman later stated, "I don't know—I don't know where Zak is. I do know that his S.I.V., his visa application is still not approved, not moved, no progress on that front. We had to work a solution outside of that system to save Zak's life, and his family's. I imagine the effort and time Major Schueman had to dedicate to arrange for Zak's flight were immense and exhausting. Leave it to a Marine to accomplish what of the U.S. Government couldn't. One can only hope that Schueman and Zak will be reunited in the near future. Zak and his family landed in Qatar on that Thursday. As they await his wife's passport paperwork, two of his children are now sick. The whole family faces a long road ahead, nevertheless relieved that they have escaped."

While Zak's situation thankfully ended in a positive manner, unfortunately, most interpreters remain fighting for their lives in Afghanistan despite the heroic efforts of many members of the military trying to assist in bringing their brother's home [82].

---

82. Michael Barbaro. "The Interpreters the U.S. Left Behind in Afghanistan. *The New York Times*. https://www.nytimes.com/2021/08/91/podcasts/the-daily/afghanistan-interepreters-special-immigrant-visas-taliban.html [Interview conducted by Lynseah Garrison & Michael Barbaro]

17. Americans Left Behind

The strong commitment of American armed forces to their fellow soldier and the country's heroes is a long-standing tradition, an unwritten policy that unites soldiers and veterans of all branches of the military. Another equally important statute recognized by all service men is the well-founded concept of "No Man Left Behind."

As American forces scrambled to evacuate Afghanistan due to the lack of any discernable exit plan, hundreds, possibly thousands of Americans were left behind. President Biden accepts responsibility for the withdrawal that he declared an extraordinary success, but how he came to this determination remains baffling. For weeks, press around the world portrayed the mayhem and often horrifying events that unfolded before the eyes of the world.

From the frantic efforts of Afghans scrambling to get into the airport to the dreadful sight of desperate people clinging to a military aircraft wheels and falling to their death, the withdrawal from Afghanistan was anything but an extraordinary success. To make things even worse, an unknown number of Americans were callously abandoned behind enemy lines, left to fend for themselves. The exact count of how many Americans remain in Afghanistan apparently cannot be accurately determined. From early estimates of 100-200 to a recent report claiming the number to be somewhere around 9,000, the number will probably never be accurately determined, but one thing that stands true is that any number more than zero is unacceptable.

Date	Estimate of how many Americans Left Behind	Stated by/Source
25 Aug. 2021	1,500	Sec. of State Blinken
31 Aug. 2021	100—200	President Biden
30 Aug. 2021	Up to 200	Sec. of State Blinken
3 November 2021	285	Dep. Sec. of State McKeon
20 September 2021	Approximately 100	Blinken (since withdrawal)
3 Feb. 2022	9,000	Senate Foreign Relations Committee

Figures in table indicate various Biden administration officials' claims (and when made) of how many Americans left behind, indicating the confusion of exactly how many Americans remain in Afghanistan

In perhaps the most accurate account to date of how many Americans were left behind in Afghanistan, a recent report released by the Senate Foreign Relations Committee discovered at least 9,000 Americans were left in Afghanistan during the withdrawal.[83] Secretary of State Anthony Blinken earlier estimated that number at closer to 100-150 and stuck to that story with minor adjustments in the numbers (See table above). U.S. Senator Jim Risch, the ranking member of the Foreign Relations Committee revealed that State Department officials believed between

---

83. Hannity, Sean. "Bombshell Report: Biden Admin Left 9K Americans in Afghanistan, Originally claimed 150." 3 Feb, 2022.

10,000 and 15,000 Americans were in Afghanistan as late as 17 August 2021. Taking in consideration the State Departments claim that approximately 6,000 people were able to escape ahead of the Taliban takeover,[84] that leaves anywhere from 4,000 to 9,000 Americans stranded.

Perhaps President Biden considered the evacuation an 'Extraordinary success' based on his statement in 2021 that, "We completed one of the biggest airlifts in history. With more than 120,000 people evacuated to safety, that number is more than double what most experts thought was possible." He went on to state, "The bottom line: 90% of Americans in Afghanistan who wanted to leave were able to leave." While an evacuation of such proportions is incredible, the fact that Americans were left behind easily overshadows the accomplishment. Could the withdrawal have been managed more efficiently? No doubt, yes, but why didn't the Biden administration utilize every tool at its fingertips to ensure all Americans were safely evacuated? A major issue was an unwavering commitment to evacuate by the symbolic date of September 11, despite overwhelming evidence that the withdrawal should have been further delayed.

Since the withdrawal, President Biden and Secretary of State Blinken have continually provided assurances that the U.S. will continue working to help Americans leave. The obvious question is how will they accomplish such a challenging feat since the evacuation "shifted from a military mission to a diplomatic mission" according to National Security Advisor, Jake Sullivan.[85] With the

---

84. According to the Daily Caller
85. Gorgey, Quint, "Pentagon: We're still obviously concerned about Americans left in Afghanistan." Politico. 31 August 2021.

hasty withdrawal from Afghanistan, America lost its military presence that could have remained in small force until all Americans, S.I.V. applicants, Afghan allies, and at risk Afghans were safely evacuated.

This sentiment was hinted at in an interview conducted by ABC News and George Stephanopoulos on August 18, 2021 in which he questioned President Biden. Here is an excerpt from that interview:

> George Stephanopoulos, ABC News, Aug. 18: So Americans should understand that troops might have to be there beyond August 31st?

Biden: No. Americans should understand that we're gonna try to get it done before August 31st.

Stephanopoulos: But if we don't, the troops will stay–

Biden: If—if we don't, we'll determine at the time who's left.

Stephanopoulos: And?

Biden: And if you're American force—if there's American citizens left, we're gonna stay to get them all out.

Despite Biden's and Blinken's assurances that the U.S. will continue to work to help Americans leave, the withdrawal of all U.S. troops before every American who wanted to leave was evacuated means Biden went back on a promise he made during the August 18, 2021 ABC News interview. Biden said at that time that U.S.

troops would stay beyond Aug. 31, if necessary.⁸⁶ Despite obvious evidence that thousands of Afghans and anywhere between two hundred and nine thousand Americans were left behind, the President went back on his statement that "*…we're gonna stay to get them all out.*" So, why didn't troops remain to see to it that all willing Americans and at-risk Afghan allies were evacuated?

In a CNBC report, the president warned that staying longer carries serious risks for allied troops and civilians. "ISIS-K…presents a growing threat to Hamid Karzai Airport." Biden went on to describe the U.S. relationship with the Taliban as "tenuous," [and continued] the longer the U.S. stays, the greater the risk that fighting will break out.⁸⁷ The whole time, fighting was already occurring as the Taliban quickly took control of city after city until finally capturing Kabul in late August.

Yes, there was a definite risk of Taliban retribution if the United States did not meet the evacuation deadline (which was twice delayed without a major spike in violence), but a military presence could have prevented abandoning thousands of Americans and allies. If the President would have stuck to his word, "*…we're gonna stay to get them all out*" and if the timeline for evacuation was adjusted, the chances of a complete withdrawal would have been possible. Instead, Americans and its Afghan allies are being hunted, persecuted, imprisoned, maimed, and in some cases, murdered on a daily basis.

Senator John Barrasso, while speaking about the fall of

---

86. Stephanopoulos, George. ABC News Interview. 18 August 2021.Full transcript of interview can be found at: https://abcnews.go.com/Politics/full-transcript-abc-news-george-stephanopoulos-interview-president/story?id=79535643

87. Macias, Amanda. "Biden says Afghanistan evacuation on track to finish by Aug. 31 deadline as threats to Kabul airport grow." 24 August 2021. https://www.cnbc.com/2021/08/24/biden-sticks-to-aug-31-afghanistan-withdrawal-deadline.html

Afghanistan in September of 2021, stated that "Afghanistan fell because President Biden paid exclusive attention to the calendar on the wall instead of the conditions on the ground...The president wanted symbolism for September 11...Well, he got symbolism but not what he wanted...It was the kind of symbolism that terrorists wanted." The Taliban recognized the president and his administration's telegraphed intent to stick to a concrete withdrawal date, further empowering them as American forces systematically withdrew from military bases throughout the country, leaving the Afghan Army to defend themselves.

Barrasso reminded the Congress that the president said 'he' (the U.S. Army) "would not leave Afghanistan and would not follow the directive that he placed, the August 31 deadline, until every American was out,"[88] still he failed to follow through on his promise and pandemonium ensued. Even as Afghan President Ghani specifically said, "Please don't start that evacuation in a major way because that in itself will lead to the collapse of our government...it will lead to a loss of confidence,...So, please don't do that," even after the Taliban closed in on Kabul and it was obvious the Afghan military was not up to the job, even as Americans and Afghan allies struggled to get to the airport through Taliban efforts to disrupt the evacuation, the president chose not to delay the evacuation again.

While addressing the Congress on the subject of Afghanistan and the withdrawal, Senator Robert Portman stated...

> "Too many people were left behind. There is no question about that. There were American citizens left behind and permanent residents left behind, and many of the Afghans who had worked with us and with our

---

88. Congressional Record S6445, 13 September 2021.

allies as interpreters, as drivers, who worked at NATO, or who worked at the United Nations were left behind—so were a lot of people who were actually in the process of getting what is called a special immigrant visa. Those would be our alles in Afghanistan..."[89]

Portman goes on to indicate that the majority of those rescued were neither American citizens, green card holders, Afghans, nor special immigrant visa holders, which reveals how the lack of vetting now places American communities at risk. Yes, the United States conducted the largest mass evacuation in history, which is admirable, but who exactly was evacuated? If most Afghans were not the tens of thousands S.I.V. applicants and holders, then who did we evacuate?

While these questions are being addressed and debated, possibly thousands of Americans are struggling to survive in Afghanistan each day. The Biden administration vows to assist those in need of evacuation using diplomatic channels. According to Secretary of State Blinken, "There is no deadline on our work to help any remaining American citizens who decide they want to leave to do so, along with the many Afghans who have stood by us over these many years and want to leave and have not been able to do so...that effort will continue every day past August 31st." These are comforting words for some, but the list of diplomatic tools are quickly diminishing as the tentacles of the Taliban take a firm hold of the country once again.

On February 22, 2022, as reported by *France 24*: "Kabul (AFP)—The Taliban will not allow any more Afghans to be evacuated until the situation improves abroad for those who have already left, their spokesman said Sunday."

---

89. See Congressional Record S7782. 4 November 2021.

Taliban spokesman, Zabihullah Mujahid declared that the promise to allow anyone to go abroad was not 'continuous,' as the Taliban quickly began to reign in on some of its more liberal travel policies that occurred in great part due to the DOHA agreement. After assuming power, the Taliban initially promised Afghans would be allowed to come and go as needed as long as they had the necessary passports and visas.

Thousands of people who served American forces, for embassies, NATO, and other organizations are still stranded in Afghanistan (along with an undetermined number of Americans) who are desperate to escape in fear of being targeted by the Taliban. According to a United Nations report, more than 100 people with links to the former Western-backed regime have been killed by the Taliban."[90] It is difficult to predict what the Taliban may do in the future. If history has anything to do with it, the quality of life for the average Afghan will continue to deteriorate, any remnants of individual freedom will be ripped from the hands of hopeful citizens, and known American collaborators and Americans stranded in Afghanistan will be persecuted, hunted, and murdered one by one.

---

90. France24. "Taliban say no more evacuations until life improves for Afghans abroad." 27 February 2022. https://www.france24.com/en/live-news/20220227-taliban-say-no-more-evacuations-until-life-improves-for-afghans-abroad

18. The Fate of the Women of Afghanistan

*"In Afghanistan, women are all heroes. They are dying
every day, but they are never giving up."*

—Laila

Senator Jeanne Shaheen addressed the Congress, specifically speaking about the fate of the women of Afghanistan following the withdrawal of United States and NATO forces. The presence of the United States led coalition forces brought much stability to the region and provided an unheard level of female autonomy in the region. Before the U.S. sent forces into Afghanistan, life for the women was severely limited. Girls were not permitted to work, go to school, and women were unable to travel without a male escort.

Things changed drastically when coalition forces, working hand-in-hand with the Afghan military, pushed the Taliban out. Afghan women enjoyed an unprecedented level of safety and security, but now all the gains are at risk. Shaheen stated, "We are leaving by September, and there is no plan to ensure that the rights that were achieved for women and girls are actually protected, even though we have legislation that says that in conflict areas like

Afghanistan, we have a responsibility to ensure that women are at the negotiating table."

In March of 2021, the State Department posthumously honored seven women who received the International Women of Courage Award.[91] Each of the seven women were murdered for choosing to challenge the exceedingly oppressive laws of the Taliban and other extremist organizations. Some of these women were assassinated for simply reporting the news, going to school, running for public office, and delivering healthcare to the needy. Standing up for basic human rights should not take an act of courage, but instead, should be a celebrated action met by reward. Instead, the courageous women of Afghanistan are now murdered and oppressed by the Taliban.

2021 INTERNATIONAL WOMEN OF COURAGE HONORARY
POSTHUMOUS GROUP AWARDEES:

Fatima Natasha Khalil | General Shamila Frough | Maryam Noorzad
Fatima Rajabi| Freshta | Malalai Maiwand |Freshta Kohistani

Senator Shaheen identified Malalai Maiwand, Fatima Khalil, and Freshta Kohistani in a picture revealing that each was murdered:

---

91. The United States Department of State document. "2021 International Women of Courage Award Ceremony." 8 March 2021. https://www.state.gov/wp-content/uploads/2021/03/2021-IWOC-Program-Book-FINAL-2.pdf

Maiwand for being a journalist; Khalil who spoke out for human rights; and Kohistani who was an activist for women's rights. The Taliban killed each of these heroic women for trying to improve the lives of other women. Senator Shaheen expressed her worry that the anticipated hostilities following coalition forces withdrawal might escalate.

She went on to provide many examples of the atrocities that occurred at the hand of the Taliban since the departure of U.S. and NATO forces:

> ...this past weekend...85 people, most of them schoolgirls, were killed in a car bomb outside of a girls school in Kabul.
>
> I saw them interviewing one young woman who, I think, was about 14, about why she thought they had been targeted. She said: "I guess it's because we want an education." This is the future we risk if we don't have a plan for how we are going to continue to support the women and girls of Afghanistan.
>
> I also want to talk about the other four women who are pictured here. Fatima Rajabi, who is in the middle, was a 23-year-old prison guard. She was on her way home from work and was on a civilian bus when the bus was stopped by the Taliban. She was kidnapped, tortured, and murdered, and two weeks later, her body was sent to her family.
>
> Then there is Freshta, who is the daughter of Amir Mohamed. She was a 35-year-old prison guard who was killed on her way to a taxi to get to work— again, killed by a gunman. At the bottom is General Sharmila Frogh. General Frogh was the head of the gender unit in the National Directorate of Security and was one of the longest serving female NDS officers in Afghanistan. She was assassinated when an IED explosion targeted her vehicle in Kabul.
>
> Finally, I think the most horrific and barbarous of all of these murders was of Maryam Noorzad. Maryam was a midwife, and she was

killed when the hospital in Kabul was attacked by the Taliban. She was there, helping a woman deliver a baby, and she refused to leave when they were attacked. She didn't want to leave the woman she was helping as a midwife, so the Taliban not only killed her when she refused to leave the woman, but they killed the mother, and they killed the baby. These are the Taliban whom weare being asked to join at the negotiating table.[92]

Senator Shaheen's worries have now come to fruition. Many women continue to be hunted and murdered along with the other Afghan allies and Americans who were left behind. Here are some brief stories of some of the few Afghan women who escaped the persecution and murderous efforts of the Taliban.

Laila, moved to the United States in 2016. She studied Islamic law and became a defense lawyer and legal advisor after attending Kabul University. When she was one or two, her father, a military general, brought the family to Pakistan because Afghanistan was at war. She returned to Kabul when she was a teenager. Her husband worked for the U.S. military—he had a construction company and they fled to America because they were both in danger. They used to go to Afghanistan once a year, but they don't now because of the Taliban. She says things are much more dangerous because of the Taliban presence.

> The people of Afghanistan—we were born in war, and we grew up in war, and we are still in war. We don't know about our future, and what happens next. Maybe it's worse. All these things are happening because many Afghan people are not educated. The terrorists, the Mujahedeen,

---

92. Shaheen, Jeanne. Congressional Record Volume 167, Number 83, Pgs. 2501-2503. 13 May 2021. https://www.govinfo.gov/content/pkg/CREC-2021-05-13/pdf/CREC-2021-05-13-pt1-PgS2501-4.pdf

the Taliban, ISIS, their mentality is not matched with educated people. But the educated Afghans, you know, they are so open minded, so good, they want everything for everyone, for every single human being. With education, at least people know their rights. I'm not giving up. I'm not silent. With my friends, I am organizing protests in Washington. And we have Afghans around the world, making protests in different countries. We are making new hashtags. This is the new thing.[93]

Laila at her home in Virginia. She worries about her family still in Afghanistan. (Credit... Valerie Plesch for The New York Times)

Hadia, who arrived in the United States in February 2020, lost two of her brothers in 1998 to the Taliban simply because they were Shiite Hazara. She was only three then, her brother thirteen.

---

93. Haridasani Gupta, Alisha & Fraancesca Donner. (Laila's statement as documented in the New York Times article: "Lost Between Border':Afghan Women on the Live They Left Behind.") 14 September 2021.

According to Hadia, they shot him in the chest and the leg and left him in the street to die. They were prevented from collecting his body. Later, in 2001, her second brother was killed in the Takhar Province and shortly after, their father died from a heart attack. Hadia thought as a result of the murder of his children since, "It was too much for him, and he just passed away by heart attack." In November of 2019, Hadia started getting anonymous phone calls and later, she noticed a car following her. Eventually, she saw the same car when she left work. The same thing happened the next day. When she informed her mom, she told her daughter, "I don't want to lose another child. You have to leave." So, Hadia came to the United States on a tourist visa in December of 2020.

Hadia at the blooming of the cherry trees in Washington in March. Credit… Valerie Plesch-NYT.

She stated that she had no choice. She arrived with a backpack, a few clothes, and her laptop…that was it. Her family went into hiding and at the time, Hadia has only spoken with her once since then for five minutes. Now the internet is down, and she has no access. All she can do is wait and hope for the best. Hadia expressed that, "Every single day I wake up with a heavy chest. I was once a role model for my generation; they saw me as someone who was helping to make a difference for them. And now look where I am. I don't even have hope for myself. I am lost—lost between borders."

Farahnaz arrived in the United States in February. A television journalist, she covered the peace negotiations in DOHA in October 2020. At that time, she interviewed Suhail Shaheen who is a spokesperson for the Taliban. Shortly after the peace talks, the Taliban began targeting and assassinating journalists. She soon discovered she was on a Taliban hit list. Two of her colleagues were already murdered. For the next few days, she stayed at home and stayed low as security forces instructed her to do. Days later, once safe, she traveled to the French Embassy and left Kabul. She watched as the Taliban went to the TOLO TV studio where she once worked. She stated, "Now the Taliban have taken over the streets of Kabul—the same streets where we, my generation, worked, protested and made music and art." She concluded by sharing that a woman's life in Afghanistan has never been easy, even over the last twenty years, but now with the Taliban in place, their lives will be much more difficult. Her final plea to the world:

> *"The women of Afghanistan do not need your sympathy, they need the world to take responsibility for the mess it created."*[94]

---

94. Alisha, Haridasani Gupta & Francesca Donner. "Lost Between Borders': Afghan Women on the Lives They Left Behind (Four Afghan women who sought refuge in America talk about their lives now and everything they gave up)." New York Times. 14 September 2021. https://www.nytimes.com/2021/09/14/us/afghanistan-women-refugees.html.

*Farahnaz looks at her clothes her friend brought back from Kabul.
(Credit... Valerie Plesch for The New York Times.)*

The toppling of the Taliban regime and the ensuing 20 years of war in Afghanistan delivered significant gains for women's rights. In less than a month after the removal of coalition forces, the Taliban and extremist groups eliminated the gains. The Taliban almost immediately closed the government ministry dedicated to women's affairs. The Taliban prevent females from returning to schools and universities. Many professional women were told to stay home until further notice, their jobs most likely eliminated. Videos show members of the Taliban flogging women in the streets, in broad daylight, for unknown transgressions.

Any signs of the Ministry of Women's Affairs Organization disappeared and were replaced by signage of the long-feared Ministry for the Propagation of Virtue and Prevention of Vice. The once busy Ministry of Women's Affairs was staffed by 90% women and worked to empower and protect the women of Afghanistan. With the elimination of the ministry, threatened women have nowhere to turn.[95]

Women in Afghanistan are now almost entirely excluded from public life. The situation is tragic. Gender equality is nonexistent and human dignity is attacked via state-sponsored oppression on a daily basis. Neither human rights nor freedom of expression exists in Afghanistan today. Many of the women in hiding, some already murdered, were unable to get to the airport for evacuation. Some were victims of the overly complicated S.I.V. application process, others simply could not get to the airport because of the unnecessary mayhem of the rapid withdrawal, but in either case, our country's allies and at-risk citizens should never have been abandoned.

---

95. Mukhtar, Ahmad. "As Taliban robs Afghan women and girls of work, school and safety, the most vulnerable 'have nowhere to go.'" CBS News. 22 September 2021. https://www.cbsnews.com/news/afghanistan-taliban-women-rights-girls-work-school-safety-post-us-withdrawal/

19. The Decision to Abandon Military Bases and Equipment

"The United States will have to deal with the fallout of this failure for years to come"

—Senator Jim Risch (Comment on Afghanistan withdrawal)

The strategic failures the United States made while withdrawing from Afghanistan will go down in history as one of the most significant national blunders ever made. From the lack of strategic planning to ensure the safe passage of Americans and allies to safety, to the flawed Special Immigrant Visa system, and the seemingly haphazard attempt to evacuate personnel during the closing hours of the withdrawal, the entire operation was acutely deficient. The United States military's many victories and the great strides made in helping to build a terrorist-free, self-governing Afghanistan, will forever be overshadowed by the fallout from the Biden administration's disastrous withdrawal.

Besides the thousands of allies and Americans left behind, billions of dollars' worth of American equipment was left for the Taliban to utilize in their terrorist efforts, but even more egregious, the mistake permits countries like China, Russia, and other adversarial countries to gain a technological advantage in future conflicts.

A U.S. Department of Defense report documents that over seven billion dollars' worth of military equipment was left in Afghanistan. The report, mandated by Congress identifies that the United States gave a total of $18.6 billion worth of equipment to the Afghan National Defense and Security Forces (ANDSF) from 2005 to August of 2021. Of that total, equipment worth $7.12 billion remained in Afghanistan after the 31 August 2021 withdrawal. The equipment left behind includes aircraft, military vehicles, weapons, air to ground munitions, communication equipment, and other materials. There are no plans to "retrieve or destroy" the equipment according to the Defense Department.[96]

The report goes on to mention that the majority of equipment used by U.S. forces was either retrograded or destroyed before the evacuation. Top U.S. defense officials have repeatedly faced criticism from some high-profile lawmakers who have called on them to account for the U.S. pullout as well as the collapse of the U.S.-backed government.

"We all witnessed a horror of the president's own making," Senator Jim Inhofe, the top Republican on the Senate Armed Services Committee, told Defense Secretary Lloyd Austin this past September, calling the way events played out "avoidable."

"Everything that happened was foreseen," Inhofe said at the time. "President Biden and his advisers didn't listen to the combat

---

96. Kaufman, Ellie. "First on CNN: US Left behind $7 billion of military equipment in Afghanistan after2021 Withdrawal, Pentagon report says." CNN. 28 April 2022. https://www.cnn.com/2022/04/27/politics/afghan-weapons-left-behind/index.html

commander. He didn't listen to Congress, and he failed to anticipate what all of us knew would happen."[97]

Many people have criticized the move to leave equipment in Afghanistan as negligent and dangerous. The equipment left in Taliban hands provides an immediate boost in their military capabilities, but in the larger picture, opens the door for even more unsettling possibilities such as permitting adversarial access to American equipment and secrets. In a *VOA* News interview, the Pentagon claimed most equipment as "…not state-of the-art stuff." The defense official, who told VOA this, spoke with them on a condition of anonymity in order to discuss the report, which at the time was not yet released to the public.

"Everything that we provided to the Afghan forces was not on the same level as ours or those of our allies" per the US official, "It was designed to fight low tech." The official also said that even the higher-end equipment was unlikely to give Taliban forces much of a boost. The high-end equipment, the aircraft, the UAS [Unmanned Aerial Systems], the precision munitions for the aircraft…that's very dependent on maintenance," the official said, noting that many of those systems "suffered very poor readiness rates" even when U.S. forces and contractors were on the ground helping Afghan forces.[98]

The unnamed U.S. official's reasoning appeared as an attempt to minimize the impact of leaving the equipment, which in the bigger picture provides American adversaries a glimpse into how the U.S.

---

97. Seldin, Jeff. "Pentagon Downplays $7B in US Military Equipment Left in Afghanistan." VOA. 28 April 2022. https://www.voanews.com/a/pentagon-downplays-7-billion-in-us-military-equipment-left-in-afghanistan/6549546.html
98. Seldin, Jeff. "Pentagon Downplays $7B in US Military Equipment Left in Afghanistan." VOA. 28 April 2022. https://www.voanews.com/a/pentagon-downplays-7-billion-in-us-military-equipment-left-in-afghanistan/6549546.html

builds and utilizes its equipment on the battlefield. Countries like China and Russia are sure to capitalize on the United States colossally foolish decision to not remove or destroy excess equipment once it became obvious the Taliban was quickly taking city after city.

American enemies will study the equipment, look for weaknesses, in order to gain an advantage in future wars. According to Patrick Tucker, a technology editor for *Defense One*, "The ultimate winner of two decades of war in Afghanistan is likely China. The aircraft and armored vehicles *left behind* when U.S. forces withdrew will give China—through their *eager partners*, the Taliban—a broad window into how the U.S. military builds and uses some of its most important tools of war. Expect the Chinese military to use this windfall to create—and export to client states—a new generation of weapons and tactics tailored to U.S. vulnerabilities, said several experts who spent years building, acquiring, and testing some of the equipment that the Taliban now controls."[99]

Removing all American equipment before withdrawing from Afghanistan clearly was out of the question. Afghan national forces needed to have some equipment to defend their country once coalition forces evacuated. At the same time, once the Afghan government collapsed and it became clear that Afghan forces were unable to stop the Taliban forces, efforts should have been made to destroy as much equipment as possible. In the end, the botched withdrawal did not permit time to do so. The equipment, now in enemy hands provides them further capabilities to spread terror throughout the region as well as provides a glimpse into some of the United States military technology secrets.

---

99. Tucker, Patrick. "How Equipment Left In Afghanistan Will Expose US Secrets: Even rendered inoperable, equipment now in the hands of the Taliban will yield troves of information about how the U.S. builds weapons and uses them." 10 September 2021. https://www.defenseone.com/technology/2021/09/how-equipment-left-afghanistan-will-expose-us-secrets/185264/

20. Why Not Delay the Withdrawal?

"No one cares... I'm thinking, every second, that the Taliban will kill my family"

—Ramish Darwishi (Former Afghan interpreter)

The DOHA agreement, whether viewed as a positive move or as a disastrous deal, existed on a series of hopes, fragile guarantees, and auspicious assurances that rested heavily on the Taliban's fear and respect of President Trump. In an interview with talk show host Sean Hannity, Trump revealed some of what was discussed with the Taliban leading up to the DOHA agreement. When asked about the very specific conditions Trump personally gave the Taliban, he eagerly shared the following:

> *I spoke on numerous occasions to the head of the Taliban, and we had a very strong conversation. I told him up front, I said, "Look, before we start, let me just tell you right now that if anything bad happens to Americans or anybody else, or if you ever come over to our land, we will hit you with a force that no country has ever been hit with before, a force so great that you won't even believe it. And your village, and we know where it is..." and I named it, "will be the first one."*

As the interview continued, Hannity spoke about the Taliban's swift takeover of Afghanistan citing that "60% of the country had already been taken over by the Taliban [as of the time of the interview (August 18, 2021). Under your plan, if they had taken over 5%, not 60, like they had last…5% that was not in your agreement, what would have happened to them?"

Taliban fighters on patrol in Jalalibad (Photograph EPA)

Trump replied, "We would have hit them very hard. Again, the words are conditions, *plural*, conditions-based. It was an agreement where, actually, we wanted to get out by May 1, and they violated the agreement. So, we didn't. It's a great agreement from a lot of different standpoints. And, frankly, Biden didn't have to even go by that agreement."[100]

Trump went on to speak about the deplorable conditions at the American / Mexican border and compared the situation at the border with what occurred in Afghanistan. His analogy basically revealed how the border was well-managed by American officials just as things were being handled in Afghanistan. Even as then President Trump reduced American forces to 2,500 in Afghanistan, a relative peace continued. Staying with Trump's analogy, all of the progress made in securing the American border during the Trump administration was completely undermined by the Biden Administration, just like Biden failed to manage the withdrawal of troops from Afghanistan. The entire withdrawal was totally mismanaged and led to Americans and Allies being trapped behind enemy lines.

As Americans and Afghan allies frantically attempted to flee Afghanistan, as the Taliban closed in on Kabul and surrounded the airport at times preventing the free-passage of personnel into the airport, as thirteen service members died in a suicide bombing outside the Kabul Airport,[101] as desperate Afghan people clung to flying aircraft tumbled from the landing gear to their deaths, at any time a halt could have been called to the evacuation. Trump delayed the withdrawal and Biden did as well, but with his eye acutely focused on a mainly symbolic date, 9/11, any chances of another delay was out of the question.

---

100. Hannity, Sean. "Donald Trump Sean Hannity Interview on Afghanistan August 17: Transcript." 18 August 2021. Thttps://www.rev.com/blog/transcripts/donald-trump-sean-hannity-interview-on-afghanistan-august-17-transcript

101. Reardon, Sophi & Eleanor Watson. "Here's what we know about the thirteen U.S. service members killed in Kabul Airport attack." CBS News. 31 August 2021. https://www.cbsnews.com/news/kabul-airport-attack-victims-united-states-military-service-members/

In a statement released by the White House Briefing Room dated August 14, 2021, President Biden blamed the former administration:

> When I came to office, I inherited a deal cut by my predecessor—which he invited the Taliban to discuss at Camp David on the eve of 9/11 of 2019—that left the Taliban in the strongest position militarily since 2001 and imposed a May 1, 2021 deadline on U.S. Forces. Shortly before he left office, he also drew U.S. Forces down to a bare minimum of 2,500. Therefore, when I became President, I faced a choice—follow through on the deal, with a brief extension to get our Forces and our allies' Forces out safely, or ramp up our presence and send more American troops to fight once again in another country's civil conflict. I was the fourth President to preside over an American troop presence in Afghanistan—two Republicans, two Democrats. I would not, and will not, pass this war onto a fifth.

It is true that the Biden Administration inherited the DOHA deal, a 1 May 2021 deadline, and that then President Trump drew forces down to 2,500, but the claim that he only had two possible choices—'follow through on the deal [or] ramp up our presence and send in more troops to fight'—is fundamentally flawed. Just days into the evacuation, the Taliban already began its overthrow of the Afghan government and in doing so, breached the confines of the peace agreement from the onset. Part II of the DOHA agreement states:

1. "…the Taliban will not allow any of its members, other individuals, or groups, including al-Qaida, to use the soil of Afghanistan to threaten the security of the United States and its allies."

2. "…the Taliban will prevent any group or individual in Afghanistan from threatening the security of the United States and its allies…"
3. "…the Taliban is committed to deal with those seeking asylum or residence in Afghanistan according to international migration law and the commitments of this agreement, so that such persons do not pose a threat to the security of the United States.

Breach #1: The security of the United States and its ally by proxy, the Afghan government forces, were threatened from the time the evacuation commenced. The Taliban marched into Afghanistan and forcefully took control of city after city until in Kabul and openly hindering the evacuation process.

Breach #2: The death of the thirteen American soldiers at the airport could have been prevented. The Taliban were tasked with preventing 'any group or individual' from threatening the United States, but thirteen American heroes were killed when a suicide bomber detonated explosives at a Kabul airport gate where U.S. troops were searching evacuees rushing to depart the country. At least 18 other troops were wounded in the bombing that killed at least 170 people as well as the 13 U.S. service members.[102] The

---

102. Scott, Andrea. "Here are the names of the 13 U.S. service members killed in Afghanistan attack." Military Times. 28 August 2021. https://www.militarytimes.com/news/your-marine-corps/2021/08/28/here-are-the-names-of-the-13-service-members-who-died-in-afghanistan-attack/

attack was the single deadliest enemy strike against U.S. forces in Afghanistan since August 2011.[103]

SERVICE MEMBERS KILLED IN KABUL AIRPORT BOMBING

- Marine Corps Lance Cpl. David Espinoza, 20, of Rio Bravo, Tex.
- Marine Corps Sgt. Nicole Gee, 23, of Roseville, Calif.
- Marine Corps Staff Sgt. Darin Taylor Hoover, 31, of Utah
- Army Staff Sgt. Ryan Knauss, 23, of Corryton, Tenn.
- Marine Corps Cpl. Hunter Lopez, 22, of Indio, Calif.
- Marine Corps Lance Cpl. Rylee McCollum, 20, Jackson, Wyo.
- Marine Corps Lance Cpl. Dylan R. Merola, 20, of Rancho Cucamonga, Calif.
- Marine Corps Lance Cpl. Kareem Nikoui, 20, of Norco, Calif.
- Marine Corps Cpl. Daegan William-Tyeler Page, 23, of Omaha
- Marine Corps Sgt. Johanny Rosario, 25, Lawrence, Mass.
- Marine Corps Cpl. Humberto Sanchez, 22, Logansport, Ind.
- Marine Corps Lance Cpl. Jared Schmitz, 20, of Wentzville, Mo.
- Navy Hospital Corpsman Max Soviak, 22, of Berlin Heights, Ohio

Breach #3: The Taliban while at first willing to allow asylum seekers to leave the country, openly hunted down Afghan allies, sometimes imprisoning and sometimes killing them in the streets. For example, a week ago, *The Guardian* offered an account of the Taliban executing a pregnant police officer in a province a bit west of Kabul:

---

103. Boburg, Shawn, Meagan Flynn, Alex Horton, et. al. "The 13 U.S. service members killed in the Kabul airport attack: The dead include 11 Marines, one soldier and one sailor, with many in their early 20s." The Washington Post. 29 August 2021. https://www.washingtonpost.com/national-security/2021/08/27/us-service-members-killed-kabul-airport-names/

Negar Masumi, a female police officer with 15 years of experience, was determined not to flee when the Taliban took control of her home province of Ghor in central Afghanistan. On Saturday night, gunmen, who called themselves Taliban mujahideen, stormed Negar's home. They took her husband and four of her sons into another room and tied them up. Then they beat Negar with their guns and shot her dead, according to a family member, who spoke on condition of anonymity for fear of retaliation. Negar, who was eight months pregnant, could not believe she would be killed because of her job.[104]

Afghans claim that the northern borders are still closed and the Taliban [are] not allowing private charter evacuation flights. Taliban spokesman, Zabihullah Mujahid recently revealed during a news conference that " … so this will be stopped [allowing people to leave the country] until we [the Taliban] get assurances that their lives [Afghan citizens] will not be endangered." He said this while responding to questions about reports that border officials had been told not to allow anyone to be evacuated—including by road.[105]

Other options existed besides the two President Biden presented to the American public. He simply revealed two possible outcomes. He chose to "Follow through on the deal." Delaying the withdrawal would have been the prudent thing to do to allow for trapped Americans and Afghan allies to be safely vetted and evacuated. Instead, the choice to stick to the deadline resulted in panic

---

104. Geraghty, Jim. "The Taliban Are Hunting Down Any Afghan Tied to Western Organizations." *National Review*. 17 September 2021. https://www.nationalreview.com/the-morning-jolt/the-taliban-are-hunting-down-any-afghan-tied-to-western-organizations/

105. "Taliban say no more evacuations until life improves for Afghans abroad." *France 24*. https://www.france24.com/en/live-news/20220227-taliban-say-no-more-evacuations-until-life-improves-for-afghans-abroad

and mayhem that led to many deaths, injuries, and U.S. citizens and Afghan allies trapped in Afghanistan.

In a scathing report, Senator Jim Risch, ranking member of the Senate Foreign Relations Committee, criticized the Biden administration for a botched withdrawal that left hundreds of Americans and tens of thousands of its Afghan partners behind. The key findings of his report are indicated below:

> The Biden Administration did not hold a senior-level interagency meeting to discuss an evacuation or formally task the State Department (State) to contact at risk populations, including Americans, until August 14, just hours before Kabul fell.
>
> The Biden Administration:
> 1. Failed to do any contingency planning for worst-case scenarios.
> 2. Ignored intelligence reports about the risk of an imminent Taliban takeover of Afghanistan.
> 3. Disregarded dissent cables from Foreign Service Officers on the front lines.
> 4. Abandoned Bagram Air Base based on arbitrary troop caps and political considerations, hampering the evacuations and the reinserted troops.
> 5. Failed to take significant steps to improve the Special Immigrant Visa (SIV) program despite clear evidence that the program was flawed.
>
> The Biden Administration failed to protect:
> - American citizens in Afghanistan—thousands of Americans and Legal Permanent Residents were left behind.

- Afghan partners—tens of thousands of SIV applicants were left behind, jeopardizing America's credibility and ability to recruit partners in the future.

The botched withdrawal damaged U.S. credibility with our allies.[106]

Risch's report comes on the cusps of what some call, the Biden Administration's willful betrayal of our friends.

In George Packard's narrative titled, *The Betrayal*, published in the *Atlantic*, he does not hold President Biden and his administration responsible for the Deal with the Taliban or them coming into power. He makes this clear in the opening of his narrative when he clarifies, "It took four presidencies for America to finish abandoning Afghanistan." However, George's narrative indicts the Biden Administration for "willful failure to offer an orderly exit from Afghanistan for combat interpreters, embassy support staff, [and] intelligence informants and so on. Many of these individuals have been denounced as traitors and are at risk of death—if they haven't been killed already."[107] Reports come out of executions and citizens disappearing on a weekly basis in Afghanistan since the withdrawal of coalition forces.

---

106. Risch, Jim. "Risch Publishes Report on Biden Administration's Strategic Failures During Afghanistan Withdrawal." 3 February 2022. *The United States Senate. Committee on Foreign Relations* (Ranking Member's Press). https://www.foreign.senate.gov/press/ranking/release/risch-publishes-report-on-biden-administrations-strategic-failures-during-afghanistan-withdrawal

107. Von Drehle, David. "George Packer's opus on Afghanistan is a scorching indictment of Biden." *The Washington Post*. 1 February 2022. https://www.washingtonpost.com/opinions/2022/02/01/george-packer-atlantic-afghanistan-scorching-indictment-bide

"The Taliban have told my family that my brothers are on a list.... They searched our house and arrested my older brother. He was released after two days, but during those days my younger brother was arrested and till now we don't know where he is, how he is, if he is alive."

— Former Afghan government official in hiding, October 9, 2021

According to Human Rights Watch, Taliban forces have summarily executed or forcibly disappeared more than 100 former police and intelligence officials since August 15, 2021. As revealed in a document titled, "No Forgiveness for People Like You," the Taliban openly hunt, arrest, and execute Afghan security forces and government officials since the American withdrawal. Many Afghans are arrested and never seen again. Family members and friends fear those taken are already dead. The report goes on to mention that the Taliban use employment records that the former government left behind to identify new targets for arrest and execution.

Baz Muhammad was employed in Kandahar by the National Directorate of Security (NDS), the former state intelligence agency. Somewhere around September 30, Taliban forces arrived at his house in Kandahar and arrested him; relatives later found his body. The murder suggests that senior officials ordered or were at a minimum, aware of the killing. The continuing executions generate fear among former government officials and others who might have believed that the Taliban takeover would bring an end to the violence.[108] Instead, many hide or wait in fear for the Taliban to arrive and take them away.

---

108. Human Rights Watch. "Afghanistan: Taliban Kill: Disappear Ex-Officials." https://www.hrw.org/news/2021/11/30/afghanistan-taliban-kill-disappear-ex-officials#

In addition to the many government officials in hiding, hundreds of interpreters, drivers, and others who worked with coalition forces face the same brutal treatment since being abandoned. Unless able to be smuggled out or if able to somehow find safe passage to freedom, many are destined to be detained, arrested, and in some cases, executed. The Taliban has taken to retaliating against family members of those in hiding.

As the Biden Administration does their best to divert attention away from those left behind in Afghanistan, many who supported America's efforts are quietly being hunted and murdered. Journalist Ken Olsen asserts, "People's lives are at stake [in Afghanistan] because they worked for the U.S. government…their whole family is in danger [and] we have a moral obligation to get them out of there.[109] The situation is dire for these people and the situation only gets worse as the days pass. Besides the growing risk of Taliban retribution, the people left behind face food shortages and a growing famine. The Biden Administration's actions or, in this case, the lack of action, is morally reprehensible and is sending a resounding message to world leaders, allies, and adversaries alike, that the United States is untrustworthy and does not care about their allies abroad.

Even as American citizens and Afghan allies attempted to withdraw from Afghanistan, the U.S. Citizenship and Immigration Services (USCIS) continued to deny thousands of humanitarian parole applications, which would permit "Afghan SIV applicants and their families to come to the United States. As of 1 July 2021, USCIS had received more than 44,500 humanitarian parole requests from Afghan nationals…as of late December 2021, the

---

109. Olsen, Ken. "Left behind: Tens of Thousands of Afghan interpreters, families are in peril as U.S. bureaucracies quibble over process." *The American Legion: Veterans Strengthening America.* (Memorial Day 2022 edition). May 2022.

agency denied 2,250 requests and conditionally approved 200. In a 5-6 month period, only a total of 2,450 requests had been decided out of the 44,500 requests. The inefficiency of the system is monumental, and, at a minimum, the current system should be torn down and rebuilt from the ground up. Peoples' lives are at stake while interdepartmental as well as government agencies needlessly argue over policy instead of doing everything humanly possible to save the lives of the Americans and allies trapped behind enemy lines.

Epilogue

"This is the greatest embarrassment in the history of our country"

—Former President, Donald Trump

"If we trust our fellow man, for him to stand up with us and to have our backs, then we can stand against the most frightening of evils."

—Douglas R. Satterfield

Many Americans share an outrage regarding the policy, or lack thereof, concerning the withdrawal from Afghanistan. A fundamental disagreement exists among Democrats and Republicans and the divisions go well beyond the political arena and spill over into the American population. Disagreements will always exist amongst people, but one concept that most if not all Americans unanimously agree upon is the belief that "No Man Should Be Left Behind."

After the last American planes departed Afghanistan, America officially shifted from a military to a diplomatic mission. A question many asked following the evacuation is what diplomatic weapons does the United States wield now that U.S. forces are gone. National Security Adviser Jake Sullivan attempted to address this question in August 2021, when he insisted that the United States held "considerable leverage over the Taliban to ensure that any remaining American citizens will be able to get out," but he never revealed the details of the supposed 'leverage.'

In reality, the most feasible way of evacuating all American citizens and Afghan allies was to do it while America had boots on the ground. Once the military departed, nearly all avenues of freeing abandoned Americans and allies was eliminated without the cooperation of Taliban officials. If history has anything to say about it, the Taliban will, as they always have, act in what is in their best interest; America's needs are not one of their concerns.

Recent numbers remain confusing as to how many Americans are trapped in Afghanistan. Most recent reports vary from anywhere between a couple of hundred to the recent Senate Foreign Relations Committee declaration that as many as 9,000 Americans were left in Afghanistan during the withdrawal. The same lack of precise numbers holds true when speaking of the many Afghan allies abandoned during the disastrous withdrawal. Early reports estimated around 150,000-250,000 Afghan allies were left behind, but more recent reports set the number at approximately 78,000.[110]

"Regardless the number, anything over zero is unacceptable."

The Biden Administration's failure to manage the withdrawal, the choice to not further delay the evacuation despite obvious

---

110. De Luce, Dan. "U.S. Left Behind 78,000 Afghan Allies in chaotic withdrawal: NGO report." *NBC News*. https://www.nbcnews.com/investigations/us-left-78000-afghan-allies-ngo-report-rcna18119

warnings that the Taliban were not going to meet the requirements of the DOHA Agreement, proved to be devastating for those who are now in hiding and who are being hunted on a daily basis in Afghanistan. But the damage goes well beyond those individuals trapped behind enemy lines. America's reputation throughout the world has been forever tainted. The stories of those that worked with America, those who put their lives at risk as they worked side by side with the United States military for a peaceful Afghanistan, those who are now cast aside and left to fend for themselves, sends a resounding message to America's allies.

Is America dependable, will they truly take care of their allies, will they meet their obligations, and do they have the resolve to follow through on their promises? Whether intended or not, the events that took place in Afghanistan will cast doubt on America's reliability and motivations worldwide. The next time America reaches out to an ally with assurances like "we have your back;" "you will be taken care of;" "we will stand side by side with you;" the doubt will be there. Thoughts of how the United States abandoned its allies and, worse than that, their own people in Afghanistan, will not be soon forgotten.

The stories have already started of employees, translators, government officials, and women's rights activists and even subcontractors who were unable to make it to the airport before the Taliban moved into Kabul. Stories of the thousands of Special Immigrant Visa applicants, human-rights activists, Afghans adopted into American military units, and many others huddling behind walls, hiding with neighbors, friends, and family, in fear that the Taliban will find them, are easy to find.

At the same time, our adversaries celebrate the stories of America turning its back on its own and Afghan allies. The kind and caring local Afghans who allied with the United States and coalition forces in hopes of a unified Afghanistan and who received assurances that they would be taken care of when the time came for withdrawal were abandoned. Countries like China and Russia will capitalize on the tales of America and the chaos they left behind in Afghanistan. They will spread the stories of how America left their own people and allies behind to face persecution and even death. They will spread stories of those 'meddling Americans' who butt into other countries' affairs with promises of peace and unification, but in the end, create mayhem, and then flee. Authoritarian regimes across the world will spread the stories of America's failure because "Doing so serves their goals [our adversaries] perfectly, discouraging dissidents from believing they have any durable protection.[111]"

As America's enemies capitalize on the catastrophe in Afghanistan, President Biden continues to deflect blame for the fiasco that he orchestrated. While at times he has openly accepted responsibility, he simultaneously blamed others, mainly Donald Trump for the failed withdrawal. At one point, President Biden informed Americans that his military commanders didn't recommend leaving troops in Afghanistan as confirmed in an August 2021, *Good Morning America* interview. When asked about his commander's recommendations, President Biden "denied that his top military commanders recommended he leave 2,500 troops in Afghanistan..."[112] Biden was described as adamantly denying "that military leaders did not argue against his plan to withdraw all troops by Sept. 11, while it was reported by *The Wall Street*

---

111. De Luce, Dan. "U.S. Left Behind 78,000 Afghan Allies in chaotic withdrawal: NGO report." *NBC News*. https://www.nbcnews.com/investigations/us-left-78000-afghan-allies-ngo-report-rcna18119

112. Chalfant, Morgan. "Biden denies military commanders recommended he leave troops in Afghanistan." 19 August 2021. *The Hill*. https://thehill.com/homenews/administration/568515

Journal that "Biden's decision to withdraw all U.S. troops from Afghanistan went against recommendations from top military commanders—specifically Gen. Frank McKenzie, commander of U.S. forces in the Middle East; Gen. Austin Miller, commander of NATO forces in Afghanistan; and Gen. Mark Milley, chairman of the Joint Chiefs of Staff—to keep 2,500 troops in Afghanistan."[113] The Biden administration has reverted to making excuses, deflecting blame, and lies for the massive failure in Afghanistan. All the many gains made over the twenty-year war, the deaths of 2,448 American service members killed in Afghanistan,[114] in the attempt to deliver peace and stability to the region were disgracefully sacrificed in a matter of a couple of months due to a series of seemingly ill-advised decisions.

Will the "No Man Left Behind" ethos endure following the calamitous withdrawal from Afghanistan? For the United States military, the creed lives on as a sacred commitment that will persist into the future. For our government, perhaps Afghanistan was an exception to the rule that will inform future leaders to the importance of fulfilling commitments and promises made to fellow Americans and allies alike.

In the meantime, the Biden administration needs to be held accountable for their willful betrayal of fellow Americans and allies abandoned in Afghanistan. The president must ensure that he follows through on his promise to, "Look at all possible options to

---

113. Gordon, Michael R. "Biden Rebuffed Commanders' Advice in Decision to Leave Afghanistan Top generals favored holding the line at 2,500 U.S. troops while negotiators pursued a peace deal." 17 April 2021. https://www.wsj.com/articles/biden-rebuffed-commanders-advice-in-decision-to-leave-afghanistan-11618696597

114. Lock, Samantha. "How Many U.S. Soldiers Died in Afghanistan?" *Newsweek*. 16 August 2021. https://www.newsweek.com/number-us-soldiers-who-died-afghanistan-war-1619685 [according to data from Linda Bilmes of Harvard University's Kennedy School and the Brown University Costs of War project, as reported by the Associated Press].

evacuate any Americans who want to come home." At the same time, the U.S. must use all diplomatic avenues to extract all willing Afghan allies. Once every American and Afghan ally has been emancipated, the United States can begin to rebuild its reputation, mend relationships with its allies, and once again become the world leader that is respected by every nation on earth.

In the meantime, our countries focus, thoughts, and prayers should be acutely focused on saving those American patriots and Afghan allies who were so callously abandoned behind enemy lines. They must NEVER be forgotten, nor should efforts to rescue them cease until they are safely evacuated because No Man, American or ally, should Ever be Left Behind.

About the Author

David Brown served in the United States Army earning the bronze star for his actions during Operation Desert Storm. After exiting the service, he earned his bachelor's degree in English and special education from the University of West Georgia and later, earned his Master of Arts degree in English and creative writing from Southern New Hampshire University. For the past ten years, he has served as a special education teacher at the Haven Academy at Alexander Comprehensive High School in Douglasville, Georgia and additionally, spent the past six years as an adjunct professor of writing at Georgia Highlands University. He and his family currently reside in Georgia.

Other Books by the Author

Stories told as they
occurred on the battlefields
of the Middle East;
many for the first time.

PTSD &
THE IMPACT OF
COVID-19
BATTLEFIELD
MANIFESTATIONS

DAVID ALLAN BROWN

KINDNESS IS ENCOURAGED

A CAUTIONARY TALE

DAVID ALLAN BROWN

STORY CONCEPTUALIZED IN THE DESERTS OF IRAQ, KUWAIT, AND SAUDI ARABIA DURING THE GULF WAR.

www.ingramcontent.com/pod-product-compliance
Lightning Source LLC
Chambersburg PA
CBHW030328100526
44592CB00010B/613